ECONOMIC MEASUREMENT OF ENVIRONMENTAL DAMAGE

A TECHNICAL HANDBOOK

by

Karl Göran Mäler and Ronald E. Wyzga

ORGANISATION FOR ECONOMIC CO-OPERATION AND DEVELOPMENT

The Organisation for Economic Co-operation and Development (OECD) was set up under a Convention signed in Paris on 14th December, 1960, which provides that the OECD shall promote policies designed:

— to achieve the highest sustainable economic growth and employment and a rising standard of living in Member countries, while maintaining financial stability, and thus to contribute to the development of the world economy;
— to contribute to sound economic expansion in Member as well as non-member countries in the process of economic development;
— to contribute to the expansion of world trade on a multilateral, non-discriminatory basis in accordance with international obligations.

The Members of OECD are Australia, Austria, Belgium, Canada, Denmark, Finland, France, the Federal Republic of Germany, Greece, Iceland, Ireland, Italy, Japan, Luxembourg, the Netherlands, New Zealand, Norway, Portugal, Spain, Sweden, Switzerland, Turkey, the United Kingdom and the United States.

* * *

TABLE OF CONTENTS

APPENDICES

NOTE BY THE SUB-COMMITTEE OF ECONOMIC EXPERTS(1): THE USE OF DAMAGE ESTIMATES AS AN ENVIRONMENTAL POLICY TOOL

THE NEED FOR MONETARY DAMAGE ESTIMATES

The development and implementation of environmental policy involves several kinds of decisions: decisions about which pollutants to control and about their permissible levels; decisions about how to control these pollutants; decisions about the siting of pollution sources and about alternative technologies with different environmental effects, etc. Obviously any prior knowledge of the consequences of these decisions would be of help for taking them in the first place. Before a decision can be made about whether a given pollutant should be controlled, it is necessary to know something about the effects of the pollutant. Before a choice between two or more alternative sites or technologies can be made, it would be helpful to have some knowledge about the potential consequences of the alternative choices.

The first step towards taking a decision is to gather all relevant information and to organise it in a manner which will aid the decision process. For example, before significant public and private projects are undertaken, the potential environmental consequences of these projects are described. The knowledge of these consequences can then be weighed against the consequences of alternative projects or actions before a decision is taken. The Environment Committee, meeting at Ministerial level, and the Council(2) have in fact encouraged the development of procedures and methodologies for forecasting and describing the environmental consequences of a project before it is undertaken.

It is desirable to describe the environmental consequences of a given action in as quantitative a manner as possible. Quantitative measures are in general more precise than qualitative measures, and the former facilitate comparisons. For this reason much of the

1) This note presents the appreciation gained by the Sub-Committee of Economic Experts of the Environment Committee, after having reviewed the draft handbook on damage estimates, on the role of damage estimates in the environmental decision process.

2) See in this regard C(74)216 - "Recommendation of the Council on the Analysis of the Environmental Consequences of Significant Public and Private Projects."

information on environmental consequences is presented in as quantitative a manner as possible. Health effects can be presented in terms of the potential number of persons affected by a specific malady. Agricultural damage may be presented as a measure of the amount of a specific crop lost. For each material suffering potential effects, damage can be given in quantitative terms specifying the amount of that material which would need to be replaced, repaired, or cleaned. Similarly, the specification of other consequences in quantitative terms will facilitate more rational decisions.

The principal drawback of this approach is most evident when the consequences of alternative sites, projects or technologies are compared. If, for example, two sites were considered for the location of a major pollution source, it would be necessary to compare all of the health effects, materials effects, agricultural effects, etc. separately, because each effect is measured in a different unit. Hence if the choice of the first site led to agricultural effects and the choice of the second decreased recreational possibilities, it would be impossible to compare these two sites because their associated losses are measured in different physical units. Moreover, in general, there are many different effects expressed in different units which are associated with each alternative, and the diversity in physical units makes it impossible to gain an appreciation of the total impact associated with each alternative.

One way to overcome this problem is to attach a monetary value to each component of damage associated with each alternative. The attachment of a monetary damage estimate to each effect allows an aggregate estimate of damage expressed in units which are readily understood by the public. Given aggregate estimates of damage, comparisons of sites, projects, and technologies can be made.

THE FEASIBILITY OF MONETARY DAMAGE ESTIMATES

The monetary damage attached to an effect would reflect the amount of money necessary to compensate those who suffered the damage. Through the use of monetary damage estimates, the different types of damage, financial losses and amenity losses can be aggregated into a single figure. Financial losses can be defined as changes in the level of expenditure resulting from environmental quality changes; amenity losses can be defined as the residual category of all other losses. The latter type of losses is sometimes referred to as psychic costs, intangible losses, or social costs. The financial losses associated with health effects, for instance, would include productivity losses due to accelerated mortality or increased pollution-induced morbidity and the increased costs of health care, as both of these categories involve changes in the level of expenditure. Amenity losses are the incremental costs over the financial losses and, in this example, amenity losses comprise the costs of suffering, bereavement and limitation imposed upon the affected individuals, their families and their society.

Experience indicates that there are two general kinds of effects to which monetary values can be attached easily. Where financial losses predominate there are relatively few obstacles to the estimation of monetary damage if sufficient data are available. The resulting estimates then reasonably reflect the true damage (or at worst, they are slight underestimates of the true damage), and they can be used with a high degree of confidence.

In addition, there is sufficient experience to indicate that acceptable estimates or underestimates of recreation benefits are achievable. The estimates can be used to help decide whether recreational facilities should be constructed or preserved and where such facilities should be constructed. For example, if a decision is to be taken on whether the pollution in a certain lake is to be abated or not, the abatement costs might be compared with the estimated benefits of abatement. Benefits in this case are defined as the damage avoided by the abated pollution and they could be the increased recreational value of the lake resulting from abatement.

The use of monetary damage estimates is not without qualification. Before these tools are applied or utilised, it is necessary to be aware of some areas of caution which are presented below.

- There are some areas where there exists little information about environmental impacts or effects. In such cases, where even physical effects are unknown, it is unrealistic to expect that monetary values can be found. For example, if the climatic effects or some health effects of air pollution are unknown or cannot be quantified, then monetary estimates are not possible.
- Monetary estimates cannot always be easily transferred from one situation to another. The physical effects may differ greatly among areas even if emission levels are the same. Hence it is not sufficient to know the total quantity of pollutants emitted; their location is very important. It is necessary to know the number of persons, materials, etc. exposed to given levels of pollution before the physical damage can be evaluated. The monetary damage to be attached to a given level of physical damage is not always transferable either. Consider a certain degree of water pollution in a lake. If the lake is a drinking-water reservoir or the only recreational resource in an area, the value of the monetary damage would be far greater than if a similar level of water pollution occurred in a lake located in a district where there are several lakes which can be utilised as recreational resources.

- There are consequences of irreversible actions for future generations, and these consequences cannot be readily incorporated into monetary damage estimates which generally reflect the preferences and values of the present generation. Hence the monetary value attached to a gorge, the flooding of which is considered as part of a hydro-electric power project, does not reflect the value of this resource to future generations. This criticism is independent, however, of the decision tool applied, as there exist no methods which can be used to reveal the preferences of future generations. The only solution is to ensure that those taking decisions involving irreversibilities are aware of them.

- There are some amenity losses that are not quantifiable in monetary terms. For example, there is no satisfactory method at present for attaching a monetary value to amenity losses associated with an increased risk of death. In these cases, it is necessary to present the effects in physical terms, and the decision-maker and the public can react to them accordingly.

USES OF MONETARY DAMAGE ESTIMATES

As the introductory example indicated, monetary damage estimates are helpful in comparing the environmental consequences of alternative sites, projects, actions or technologies. Related to this use are decisions as to whether or not an action should be taken. In these cases, the consequences of undertaking or not undertaking the proposed action can be compared through the use of monetary damage estimates.

In addition, monetary damage estimates can be of use in setting environmental standards. The costs of achieving a given standard can be estimated along with the costs of the damage which would occur at the environmental quality level determined by the standard. The optimum standard would be that environmental quality level where the total costs, control costs and damage costs, are minimised.

Monetary damage estimates can be used in the determination of pollution charges where these are used as instruments to achieve environmental policy goals. Monetary estimates of damage have been used in compensation schemes where victims have received payment for the damage they have suffered. Finally, these estimates of damage can give some idea of the relative magnitude of a problem in a given region or nation.

In these applications monetary estimates of environmental damage can be useful even if they are not precise. If some of the components in a damage estimate are biased and the biases are all in the same direction, the resulting estimate is known to be an underestimate or an overestimate and can be as useful as a precise estimate. If, for example, a project which will have environmental benefits is under consideration, and it is desired to compare these environmental benefits with the costs of undertaking the project, a known underestimate would be as valuable as a precise estimate of benefit (damage), if the former were greater than the costs of undertaking the project. Similarly, if an overestimate were less than the costs, it would contribute as much information as a precise estimate.

Precise estimates of monetary damage are not always necessary when they are used to compare the environmental impacts of alternative locations, projects, processes or technologies. The monetary components of damage are likely to be biased in the same direction for each alternative, and for this reason a comparison of the total monetary damage associated with each alternative can nevertheless be instructive. In addition, in those cases where lack of information prevents the attaching of an estimate of monetary damage to all components of damage, the application of this tool remains useful. Because the total number of impacts is often very large, the partial aggregation of impacts through the use of monetary damage estimates reduces the quantity of information present in any necessary comparisons. It would still be necessary to compare some impacts in qualitative terms and others in various quantitative terms, but the total number of impacts to be compared is reduced when impacts which can be monetised are aggregated into a single number.

CONCLUSIONS

Monetary damage estimates are one useful tool in the development and implementation of environmental policy. They should not be regarded as an exclusive or perfect tool but, in the absence of a perfect tool, they are a useful way of organising and summarising information on environmental effects in a way which can facilitate more rational decisions. Moreover, even when monetary estimates of damage are not precise, they may remain useful and their use should be encouraged.

There is a need for more knowledge about the quantitative effects of changes in environmental quality. Current environmental policy is often based on effects known only in a qualitative manner or with a high degree of uncertainty. Increased knowledge about the quantitative effects of given environmental quality changes would lead to improved environmental policy. Such knowledge will also make it possible to obtain monetary estimates of damage which, in turn, can be applied to help develop environmental policy. In the past it has been this lack of knowledge about quantitative effects which has limited the use of monetary damage estimates.

There is a need for further research to develop methods to estimate the monetary damage associated with losses of amenity. For some amenity losses, as losses in recreational opportunities, a methodology is well-defined and widely accepted. The situation with respect to some other amenity losses is, however, less fortunate as good monetary estimates are not as yet obtainable. There continues to be a substantial research effort applied to the estimation of amenity losses, and it may be necessary to re-examine the topic of monetary damage estimates periodically to keep abreast of recent progress made in estimating and in using them.

ACKNOWLEDGEMENTS

The preparation of this handbook was greatly facilitated by the helpful comments of a host of colleagues and friends. It would be impossible to list all of those who have helped us, but a list of those whom we wish to thank for their advice and comments would include the following: Members of the Staff of the OECD Environmental Directorate, Alain Chambournier, Tom Waddell, Don Gillette, David Pearce, Hans Opschoor, Walter Barber, Martin Singer, Peter Reynolds, Yoshitake Ota, John Richards, Edith Rosenblum, Eric Jabouille La Salle, and the participants of the 1972 OECD Summer Seminar on damage functions. In addition, the context of the manuscript has been presented at seminars at the Resources for the Future Inc., the Stockholm School of Economics and the Center for Operations Research and Econometrics, University of Louvain, Belgium. We would like to thank the numerous participants of these seminars for their useful suggestions.

One of the authors has had an appointment at the Economic Research Institute at the Stockholm School of Economics during which period he has contributed to the content of the handbook. At the Research Institute, he has been working in the general area of environmental economics with financial support from the Riksbankens Jubileumsfond (Bank of Sweden Tercentenary Fund) and some of the output of this work is reflected in the handbook.

To these people and organisations and to many others, we wish to express our sincere gratitude. At the same time, however, we assume full responsibility for any inaccuracies, errors or misjudgments which may be present in the text.

Karl-Göran Mäler
Ronald Edward Wyzga

PREFACE

This handbook is an outgrowth of the work of the Sub-Committee of Economic Experts of the Environment Committee. The former group charged the OECD Secretariat with investigating the potential utility of monetary damage functions as a tool in developing and executing environmental policy. The first step in this task was the organisation of a summer seminar(1) which reviewed some of the foundations, methods and uses of environmental damage functions and which suggested a work programme for the OECD in this area. Following upon one of the suggestions of the seminar participants, the Sub-Committee of Economic Experts asked that a handbook be prepared on the subject, the result of which is the present monograph.

The objectives of this monograph are manifold. It is designed to acquaint those who develop environmental policy with the concept and utility of environmental damage functions. The handbook indicates those situations where monetary damage functions can be applied usefully in the development and execution of environmental policy. This publication reviews and suggests methods which can be used to estimate environmental damage functions and it provides guidelines to ensure that the resulting estimates are meaningful and in accord with economic principles. A handbook can also lead to greater harmonization in the types of studies undertaken, thus allowing some studies to make use of the results of others. Moreover, consistent and well-founded monetary estimates can be more fairly compared with other estimates, be it in a benefit-cost analysis or in an attempt to weigh the merits or disbenefits of several alternatives in an allocation problem. Finally, the existence of a review of the state-of-the-art of the estimation and utilisations of monetary damage estimates can help stimulate further research in those areas where it is needed.

1) The papers presented at this seminar are published in the volume, Environmental Damage Costs, OECD, Paris, 1975.

Chapter I

INTRODUCTION

A. DEFINITION AND OVERALL ASSESSMENT

An environmental damage function is an estimate of the relationship between given levels of environmental quality and monetary estimates of the environmental damage associated with these quality levels. Hence, for a given air quality, the appropriate environmental damage function would estimate the value of the damage to health, vegetation, materials, etc. resulting from the given air quality. The idea of a damage function is not a new one: similar estimates have been constructed for generations and the concept has been widely applied to the area of environment in the last decade. The lack of a conceptual framework associated with many of the past attempts to construct environmental damage functions has contributed to their misuse and inaccuracy. The examination of the estimation, utility and limitations of environmental damage functions defined with respect to a conceptual framework will hopefully, contribute to their wider and wider use in the future.

The conceptual framework utilised in this monograph is that of standard neo-classical welfare economics and assumes that the damage to be estimated is that perceived by those who currently suffer the damage. The monetary damage is defined as the least amount necessary to compensate those who suffer from a change in environmental quality; in economic jargon, this damage is called the compensating variation. There are two major criticisms of this framework: it accepts the existing income distribution and it ignores the preferences of future generations. The response to the first criticism is twofold: it is unclear whether environmental policy should be an instrument for income redistribution, and if it is, once the redistribution goals are defined, the procedures presented below can be altered through appropriately weighting the components of the damage(1). The second criticism has no ready answer, except that this problem is not specific to environmental damage functions

1) See Chapter III.

and the preferences of future generations are unknown. The fact that these preferences may be different from existing ones should be made explicit to those responsible for developing environmental policy.

Environmental damage functions are not to be considered as the ultimate tool which will dictate or indicate the policy path to be followed. They are but one input among several which may facilitate the development of an environmental policy.

At present, no estimate of an environmental damage function is perfect. Most often estimates of damage are available for only one or two poorly defined levels of environmental quality. This imperfect situation is often due to the fact that the knowledge and measure of physical damage associated with a given environmental quality is unknown or highly speculative. In addition, there can be difficulties in attaching monetary values or price tags to physical damage.

Despite these shortcomings, environmental damage functions can be beneficial in developing environmental policy, and whenever a major intervention in the environment is contemplated those making the final decision should consider attempts to estimate the monetary damage tied to the intervention among other factors. At most, the monetary damage estimate can help provide a rational basis for a policy choice, even if the monetary estimate is far from perfect. As indicated below, imperfect estimates, particularly underestimates, can in some circumstances be as valuable as perfect ones. At worst, attempts to estimate monetary damage may force some additional degree of rationality to the decision-making process and may yield some additional information which will aid this process. Such attempts can also indicate those priority research areas where new information is needed.

B. USES OF ENVIRONMENTAL DAMAGE FUNCTIONS

If benefit-cost analysis are to be undertaken in the environmental domain, then some measure of monetary damage (benefit) is necessary. A classical example in the use of benefit-cost analysis would be that in which the (marginal) costs of pollution control are compared with the (marginal) benefits. Such a comparison could indicate whether or not a project, such as the cleansing of a polluted lake, should be undertaken or it could indicate the level at which some environmental standard should be set(1). The lack of

1) Economic theory indicates that the optimum level would be that point at which the marginal benefits would equal the marginal costs of control.

good estimates of damage or benefits can prevent the undertaking of a benefit-cost analysis, but precise estimates of monetary damage are not always necessary. For planning purposes, for example, rough estimates are often acceptable. When the difference between the benefits and costs is large, grosser estimates of the benefits can be tolerated. Moreover, when the benefits are underestimated, as is often the case, precise estimates are not necessary if the underestimated benefits are greater than the estimated costs.

Precise estimates of monetary damage may not be necessary when comparing the environmental effects of alternatives. Examples of such comparisons may involve a choice among several locations in siting an industry or the choice of an energy generating system. In such cases the damage components in the monetary estimates are likely to be similar, hence they are likely to be biased in the same direction. A consistent bias can allow a reasonable analysis even when the estimates are approximate.

Another advantage of this application of damage estimates is that it may be partial. If one or more components of the environmental damage cannot be evaluated in monetary terms, then a vector of damage can be presented for each alternative, some components of the vectors in monetary terms and others in physical or qualitative terms. The vectors can then be compared.

Imprecise estimates of damage may also be sufficiently adequate to help indicate the magnitude of environmental problems in a given nation or region. Such estimates can also be used to help determine national or regional priorities in allocation problems. Related to this use is that which several governments have already made of monetary damage estimates where they serve to reinforce governmental environment policy.

There are other uses of environmental damage functions such as bases for setting charges on emissions to the environment or as inputs to determine awards in litigation procedures. In general, however, these uses require more precision of damage estimates than is usually possible to achieve at present.

C. SCOPE OF THE HANDBOOK

The content of the handbook is divided into seven chapters and covers most aspects which are necessary to understand, to estimate and to use environmental damage functions. The second chapter provides an explanation of the link between the environment and monetary damage functions and introduces the concept of environmental damage functions in an informal manner. The following chapter presents the fundamentals of economic theory necessary for a proper understanding of environmental damage functions and as such, this chapter presents them in terms of the conceptual framework of welfare economics. The presence of a conceptual framework facilitates the discussion of the methodology used to estimate environmental damage and of their use. In fact, it would be impossible to consider damage functions without some conceptual framework. Within this framework, financial losses are defined as changes in expenditure due to environmental quality changes, while losses of amenity are roughly defined as the residual monetary damage over and above the financial losses. The fourth chapter distinguishes between these two general categories of losses, and it presents a list of the major types of monetary damage which can occur in each category.

The next two chapters focus upon methods which can be used to estimate monetary damage functions. Chapter five presents and illustrates those methods which can be used principally to estimate financial losses. As financial losses are defined as changes in expenditure due to changes in environmental quality, the estimatation of these losses is, in principle, straightforward. The physical damage functions give the physical effects of environmental changes. If the physical effects on materials, for example, are known, it is possible to compute the financial loss with a knowledge of the costs of replacing or protecting the damaged materials. Similarly, given the necessary information on physical effects, other financial losses can generally be calculated. There are often serious accounting problems present in these calculations, however, and these and other problems associated with the estimation of financial losses are discussed in chapter five.

Chapter six concentrates on those methods which can be used to estimate losses of amenity. In contrast to financial losses, losses of amenity can create serious theoretical problems. Because these losses are not directly observable, more sophisticated methods have to be used to estimate them. One natural approach is to ask the individuals concerned through the use of a questionnaire about the necessary compensation. Another method is to try to find some connection between the loss of a specific amenity and the demand for a private good. With such a relation it may be possible to deduce an estimate of the amenity loss. Sometimes these losses are capitalised in lower land values and a comparison of land values, either over time or between different districts may yield an estimate of a loss of amenity. Finally, litigation results may provide an approximation of amenity losses.

The final chapter presents some recommendations and conclusions. The choice of a method is discussed as a function of the type of damage to be estimated, the availability of information and the purpose of the study. Some suggestions are made as to the most appropriate methods for given circumstances. This chapter also realistically evaluates the potential applications of environmental damage functions, indicating those occasions where their use and value is strictly limited and others where their use is most helpful.

Chapter II

Damage functions and the environment

A. SUMMARY

The purpose of this chapter is to provide a non-theoretical explanation of the link between the environment and monetary damage functions.

As such, it is designed to provide non-environmental economists with an introduction to the environmental context to which damage functions must be adapted. This chapter also exposes non-economists to the neo-classical approach of economists in attacking environmental problems.

A number of concepts are defined in this chapter; the most important definitions are as follows:

The environmental interaction function is defined as the relationship between emissions or residuals and the resulting ambient quality;

A physical damage function is the relationship between changes in the ambient environmental quality and physical effects on health, materials, productivity, etc.;

The monetary damage is the amount necessary to compensate those who suffer from a change in environmental quality;

The abatement costs are those costs incurred in order to achieve a change in the ambient environmental quality;

The damage cost is the monetary damage which results from the physical effects of changes in the ambient environmental quality.

B. THE ENVIRONMENTAL AND ECONOMIC CONTEXT

Environmental damage functions are used to evaluate environmental changes resulting from human activities. This tool facilitates comparison between benefits and costs in environmental policy, determination of priorities, setting of appropriate charges, and provision for a total measure of environmental degradation.

In order to obtain a clear understanding of environmental damage functions, it is necessary to have some general frame of reference, both with respect to the environmental context and to the functioning of the economy.

The environment is usually defined as the aggregate of surrounding things, conditions and influences. For our purposes, we limit our consideration to the natural environment which consists of the air mantle, the rivers, lakes and oceans, and the land, together with the biological life in these environmental media. The concept is relative to the geographical boundaries which determine the social system under study.

The environment provides various services to the social system, and for this study it is useful to classify these services into three main categories. The first, materials-input services, consists, for example, of raw materials in the form of minerals, agricultural products, space for buildings and roads. In principle, these services can be privately appropriated and markets for them can be established. (This is, in fact, the case for most of these kinds of services in the OECD Member countries.)

The second category consists of the waste disposal services which the environment can provide. In general, there are no markets or prices for these services although it is possible to establish markets or prices through government intervention.

The third category is labelled environmental services and includes a variety of services, such as the life-supporting and amenity services. The supply of these services depends on the ambient environmental quality.

All of these environmental services are interrelated in a very complex way. An increase in the use of the waste disposal services generally reduces the supply and quality of other environmental services. This is probably also true for increases in the use of the materials-input services. This relation, in which the ambient environmental quality is a function of the use of materials-input services and the use of the waste disposal services, is defined as the environmental interaction function.

In order to simplify the presentation, we neglect the materials-input services. The situation can then be illustrated in Diagram 1.

All economic activities generate by-products. Some do not have a sufficiently high market value and it is not profitable to process such by-products further. They are therefore discharged into the environment. Such by-products are denoted residuals or wastes. This classification of by-products into residuals and non-residuals is dependent upon the set of existing relative prices. At a different set of prices, it may be profitable to recover and

Diagram 1

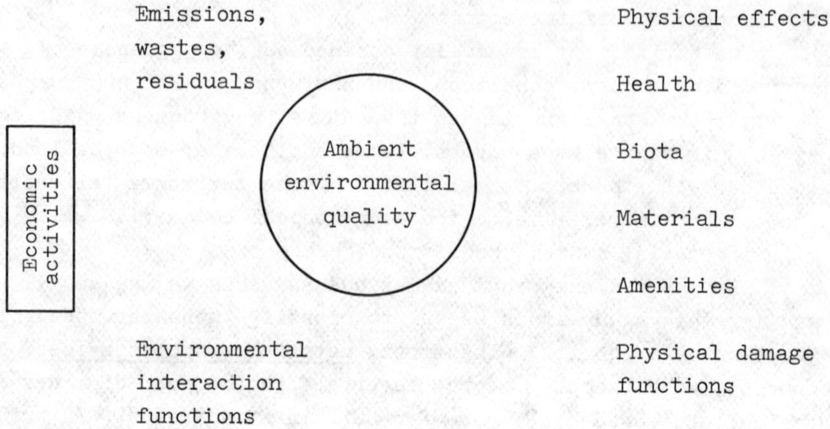

reuse some of the by-products that are presently classified as residuals. Moreover, at one set of relative prices it may be profitable to discharge a certain substance into a water body, while at another set of relative prices, it may be beneficial to the decision-maker (e.g. a householder or a plant manager) to release the same substance into the atmosphere. Changes in relative prices can also provide incentives to change the form of residuals; e.g. from a hazardous form to a harmless form.

The discharge of residuals into the environment causes changes in environmental quality. Throughout this monograph we will assume that it is possible to define and measure ambient environmental quality. A general and all-purpose definition of ambient quality is impossible. The definition must be appropriate to the problem under scrutiny. Examples of this concept include the concentrations of sulphur dioxide in the atmosphere at certain grid points, the dissolved oxygen contents at given locations in a river, the existing number of a certain species.

The ambient environmental quality is a function of the emissions of residuals into the environment. This relationship is called the environmental interaction function. It is obviously very simple: the relationship between emissions and environmental quality is dependent upon complicated meteorological, geographical, hydrological and ecological mechanisms. Moreover, some of the parameters which influence these mechanisms are of a stochastic nature. The concentration of sulphur dioxide, for example, depends not only upon the emissions, but also upon the prevailing winds, precipitation, etc.

As weather can be considered a random variable, the concentration of sulphur dioxide is also a random variable. Hence, in the consideration of environmental questions, the knowledge of the complete probability distribution of this concentration is desirable. In practice, it may be necessary to use some statistics which help to describe the distribution, such as the expected value, the upper decile, etc. We will discuss the consequences which these stochastic aspects may have on monetary damage functions later.

Changes in ambient quality can cause direct physical effects, such as those on health, corrosion, reduced agricultural productivity and aesthetic deterioration. These effects are the <u>damage</u> or the <u>physical damage</u>. The relationship between ambient quality and physical damage is the <u>physical damage function</u>. The principal consideration of this monograph is the allocation of a monetary value to the physical damage.

So far, we have introduced a classification system on which a discussion of environmental policy could be based. In order to go further, it is necessary to introduce assumptions about the functioning of the economy. To facilitate the following discussion, the assumption is made that the economy is perfectly competitive and that it has no imperfections except those explicitly related to the environment. This assumption need not reflect the real world, but must be regarded as a device which facilitates discussion of the environmental problems which follow. It implies that all economic agents are optimally adjusted to prices determined so that demand equals supply. Economic theory tells us that if we can disregard the environmental effects, the equilibrium is such that resources are allocated in such a way that there is no alternative feasible allocation in which no one is worse off and at least one individual is better off. In such a situation, the allocation is called <u>Pareto optimum</u>.

The presence of environmental effects prevents the attainment of such an optimum through the usual market mechanism. For example, the discharge of wastes changes the flow of environmental services, which in turn affects the welfare of individuals as well as the production possibilities of the firm. These effects are not taken into account by the price system.

Let us assume, for example, that a firm, as a result of innovation, increases its production and its discharge of particulates into the atmosphere. The increased ambient concentration of particulates will have several effects noticeable to the individuals living in the neighbourhood of the firm, such as dirty windows, an unpleasant atmosphere and perhaps health effects. These effects reduce the consumers' welfare and are not reflected in the prices of the firm's products. This situation is hardly a Pareto optimum. The individuals suffering from pollution would be willing to reduce their materials consumption if this would result in a better air quality. If the reductions in consumption could be transferred to the polluter as a compensation for a reduction in pollution, no one would be worse off and at least one individual would be better off. There then exists an alternative feasible allocation of resources in which one individual is better off and no one is worse off. It might be inferred from this example that pollutees should compensate the polluter and not vice versa. This is not the case, and in chapter III we will consider this and similar equity problems in more detail.

The main objective of this study is the investigation of methods which can be used to attach monetary values to physical damage. With such a monetary value it would be possible, for example, to see whether the benefits of reduced emission levels would outweigh the costs of abating these emissions; that is, one could undertake a benefit-cost analysis of such abatement.

One could try to measure the monetary value of the reduced welfare by measuring the increased expenditure for cleaning, air conditioning and medical care, but this expenditure is not, at least in principle, a correct measure to be used in a benefit-cost analysis. The aim is a number which represents the individual's own evaluation of his loss in welfare. This leads us to the following provisional definition of monetary damage: "the monetary damage to one individual resulting from a change in environmental quality is the amount which would be necessary to compensate that individual so that he is not in a worse situation because of the change in environmental quality".

This definition also covers the case in which changes in environmental quality alter the probabilities of certain effects. In this case the monetary damage to the individual is the amount necessary to compensate him so that he is indifferent to the change in the probability of a given outcome.

Hence, as an increase in the concentration of particulates leads to increased soiling, the monetary damage to an individual is the amount of compensation he must receive for his expected increase in expenditure on cleaning, as well as for his expected increase in psychic costs, or increased loss in amenity. If, in this example, the amount needed to compensate the damaged individual were greater than the abatement costs, then the choice would be for abatement.

In considering a pollution abatement programme which would improve the quality of the environment and consequently reduce the monetary damage, we could equally well use the concept of benefits from the programme. In this chapter, benefits from an improvement in the quality of the environment are loosely[1] defined as the corresponding reductions of the monetary damage. As we are now considering an improvement in welfare, the compensation used in the definition of monetary damage will be negative, and is, in fact, the maximum amount the individual is willing to pay to obtain the results (benefits) from the abatement programme. The underlying idea in the definitions of benefit and damage are the same: how much do we have to transfer to (in the case of a reduction in environmental quality) or from (in the case of a quality improvement) the individual so that he accepts a change in environmental quality? This amount can then be called monetary damage or benefit, depending on the sign (+ or -); one concept being the opposite of the other.

These definitions only indicate the magnitude of the necessary compensation; they do not require that this compensation be paid. As long as we discuss only one individual, this does not create any problems, but in chapter III when we define monetary damage for a group of individuals, it is necessary to elaborate upon this point. If the increased concentration of particulates affects the production possibilities of a firm, then this firm will experience a loss in profit. If prices do not change as a result of the damage, the monetary damage, the necessary compensation is equal to the loss in profit. If many firms are affected by the pollution, then prices will probably change and the damage will diffuse into the economy. The above definition can still be applied, but the empirical problem of estimating the monetary damage is now enormous. If these price changes are small, they can be ignored.

Aggregate or total monetary damage is provisionally defined as the sum of individual monetary damage. This definition raises a number of problems connected with income distribution, which will be discussed in chapter III.

Let us illustrate these definitions through an example. Let us assume that a firm increases its waste load to a river because this would increase its profit. Assume also that it is possible to estimate the aggregate monetary damage due to the increased pollution. If the increase in profit is greater than the monetary damage, then it is possible to redistribute the increase in the profit in such a way that no one is worse off and at least one individual is better off. If such redistributions are possible, then our definition of monetary damage works perfectly in such a benefit-cost analysis, and in fact, the definition corresponds directly to those of prices on goods and services. If, however,

---

1) This definition will be made more precise in chapter III.

such transfers are not feasible, some explicit value judgment is necessary. This point will be elaborated upon in chapter III.

The approach followed here is the classical approach in benefit-cost analysis, and is based upon the individual's evaluation of the damage. In some cases this approach may not be fruitful because of the failure of the average individual to perceive certain changes in environmental quality. An increase in air pollution, for example, may have some observable effects such as dirtier windows, eye irritation and foul odours. The individual has the capacity to perceive and comprehend these effects but there may be other effects which are not as easily perceived by the individual, such as long-term physiological effects. If these health effects are not perceived by individuals, one can perhaps argue that there are no costs associated with the effects. On the other hand, in the long-run, the individuals will be aware of these effects (especially those who will suffer from pollution-induced illnesses) and more resources will be allocated to medical care so that there are real costs from the air pollution. One solution to this problem would be to inform individuals about all conceivable effects from air pollution and then base estimates of monetary damage on the previous definition. Education, is of course, an important aspect of environmental policy, but from the point of view of damage functions, it is not a solution because we need estimates of such functions long before it can be assumed that all individuals have sufficient information. Moreover, some of the methods used to estimate damage functions are based on time series data which do not reflect changes in perception. We must therefore look for some other solution. One approach is to base the definition on the perceptions of an informed community, i.e. experts, and then try to evaluate these perceptions in monetary terms through use of some set of 'reasonable' prices. Although these expert perceptions may be more 'objective', any monetising of the environment must include subjective evaluations and it is in no way clear that the opinion of an informed group will mirror the values the general public would have had if the latter were as well-informed as the expert group.

The environmental interaction function has been defined as the relationship between, on the one hand, ambient environmental equality, and on the other hand, the use of the environment's waste disposal services (i.e. emissions). We could also include the use of materials-input services in this function in order to be able to discuss the effects arising from the extraction of raw materials. The definition of environmental quality was seen to pose problems and is dependent upon a specific situation. In most cases environmental quality will refer to ambient concentrations of pollutants, but in other cases, the quantity of discharged wastes may be taken as a measure of ambient quality.

We recall that the physical damage function is the relationship between the physical damage and ambient quality. It is this physical damage that is of the greatest interest when discussing monetary damage. The physical damage is related to the use of the waste disposal services of the environment via the ambient qualities. We examine the relationship between emissions and physical damage in two separate steps for reasons of convenience. The division between the steps is quite arbitrary. The basic idea is that "the costs of measures taken to change the ambient quality are pollution abatement costs, while the costs of those measures taken to reduce the physical damage leaving the ambient quality unchanged is part of the monetary damage". For example, in the case of aircraft-generated noise nuisance, if the ambient quality is defined in terms of the noise emitted from the aircraft, the costs of retrofitting the aircraft are abatement costs; whereas, those costs incurred in double-glazing windows to reduce noise penetration into homes are included in the damage costs.

We will consistently assume that individuals and firms adjust to a change in environmental quality in an optimal way; i.e. if a firm notices that the increased ambient concentration of a particular pollutant reduces the life expectancy of a material it uses, it will counteract this physical damage in a way which will minimise its total cost (i.e. the cost of counteracting measures, plus the residual monetary damage).

The assumption of optimal behaviour by the firms and individuals suffering from physical damage is a very important one, because without this assumption, it would not be possible to define a unique monetary measure of the damage. In general, there exist many different ways through which individuals and firms can avoid some of the physical effects from decreased environmental quality; we therefore need a principle with which we can describe their approach. The only non-arbitrary principle is the assumption of optimal behaviour, i.e. firms and individuals choose the alternative that minimises their costs.

C. EXAMPLE

The above ideas will now be applied to the case of sulphur dioxide emissions. The direct result of these emissions is increased concentration of sulphur dioxide in the atmosphere close to the source, and it is usual to define the environmental quality in this area as the ambient concentration. High ambient concentrations of sulphur dioxide will in general lead to effects on health, plant life and materials.

In order to reduce this physical damage, it is possible to take measures which increase environmental quality or which reduce the physical damage from a given ambient concentration of sulphur dioxide. Among the first measures are reduced emissions and the construction of taller smoke stacks, which disperse the sources over a greater area. The costs of these measures are control costs or abatement costs. The costs of the physical damage are a component of the monetary damage from the emissions.

Let us now consider a firm whose materials are corroded by the sulphur dioxide it emits, and let us assume for the moment that these are the only effects of these emissions. The firm can react in several ways: it can replace the corroded materials with more resistent materials; it can move to a less polluted area; or it can accept a shorter economic life for the materials. The assumption of optimal behaviour means that the firm maximises its profit and will react in the way that reduces its profit least. The cost to the firm in this case is its reduction in profit; this is the monetary damage as defined in the preceding section.

In the case of an individual who suffers a respiratory disease as a result of the increased concentration of sulphur dioxide, the reasoning is slightly more complicated. The individual can move to a less polluted area, he can stay put and try to avoid the disease by using air conditioning equipment, or he can stay put and suffer the disease, paying the extra medical expenditures and perhaps losing his wage income while he is sick. With each of these choices, it is possible, in principle, to associate the compensation necessary to the individual so that he does not feel worse off. Our assumption of optimal behaviour implies that it is the smallest of these compensations which defines the monetary damage.

Sulphur dioxide emissions can also affect the acidification of the soil, lakes and precipitation(1).

Sulphur dioxide can be transported over long distances, and during this journey it is oxidised into sulphuric acid. The sulphur is removed from the atmosphere through a) rainfall or snowfall, b) the precipitation of sulphur-containing particles to the ground, and c) the absorption of sulphur dioxide by vegetation. In some soils the deposition of sulphur and acid will lead to soil acidification with negative effects on productivity in agriculture

1) Royal Ministry for Foreign Affairs and Royal Ministry of Agriculture (1971) Air Pollution Across National Boundaries: The Impact on the Environment of Sulphur in Air and Precipitation, Sweden's Case Study for the United Nations Conference on the Human Environment, Stockholm.

and forestry. It may also lead to acidification of some water bodies. This, if sufficiently serious, can lead to the extinction of life in the water bodies.

As the ambient concentration of sulphur dioxide may be independent of the deposition of sulphur, the former may not provide a good definition of environmental quality. Good measurements of the latter unfortunately do not exist. As the amount of sulphur deposited over a wide area could be expected to correlate highly with the total emissions of the area for certain regions, these latter serve as a better definition of environmental quality than measures of ambient concentration when soil or lake acidification is of interest. This example indicates that there is no unique measure of environmental quality. In the case of sulphur dioxide at least two measures can be of value.

In general, environmental quality is a multi-dimensional concept, and the variables which describe environmental quality are inter-related in a complex way. "Hence, in order to make a proper choice of a sub-set of these variables as a basis for the estimation of damage functions, the particular context and decision situation must be specified."

In some cases special problems arise in defining the physical damage function and the monetary damage. Let us examine, for instance, the case where the deposition of sulphur and acid reduces agricultural productivity. Assume that there is an area of land which has not been cultivated because of a low price level for agricultural products. There is therefore no monetary damage associated with the soil acidification in this area. Agricultural prices increase as a consequence, and portions of this parcel of land are cultivated. What is the monetary damage resulting from the deposition of sulphur and acid on this area of land? It is difficult to answer this question because allocation issues are intermingled with equity issues. It is clear that from the social point of view some reduction in sulphur emissions is desirable. If the pollution (soil acidification) is abated and the costs of abatement are borne by the farmers, agricultural profitability will diminish, and the profits foregone because of the pollution (soil acidification) will consequently be small. If, on the other hand, the polluter-pays principle is applied, the abatement costs will be borne by the polluters and agricultural profitability will be higher. In the first case, monetary damage will be comparatively small, while in the second case the monetary damage will be large.

Different distributions of pollution abatement costs thus have different effects on agricultural profitability and therefore also on the monetary damage caused by pollution. "In some cases it is therefore necessary to state explicitly some set of value judgments as to how the burden of pollution control will be distributed between the polluters and the victims" before one can properly define damage. More generally, the distribution of wealth may substantially affect the monetary damage.

One further difficulty must be recognised. Even if we have succeeded in defining the materials-input services, the waste disposal services and the flow of environmental services, there remains the problem of estimating the environmental interaction function and the physical damage function. It was already concluded in the previous section that this function is not a deterministic relation, but a function which provides probability distributions of environmental quality for a given set of materials-input services and waste disposal services.

In many cases it is not sufficient to estimate the damage as a function of the ambient concentration because this concentration may vary considerably during a given period. For example, it may be that the damage from a moderate but permanent increase in the concentration of SO_2 is not particularly great, but that a single episode of high SO_2 pollution may create substantial damage, despite the fact that the mean concentration remains relatively low. "It is therefore not sufficient in many cases to base an estimate of the monetary damage on the average value of the environmental quality variable. In these cases, one needs some knowledge about the probability distribution of the environmental quality before one can estimate the monetary damage."

The stochastic nature of events also applies to the physical effects resulting from an environmental change. For example, if the air quality were to decrease, it might be possible to estimate the resulting increase in respiratory disease cases, but it is not possible to predict which individuals will be affected. One can only predict the increase in the probability of contracting the disease.

Finally, there is the problem of incomplete scientific knowledge. We do not know the complete interaction function and the physical damage function with any accuracy. For some effects, for example, the effect of organic waste loads on oxygen content in a river, we have a fairly good knowledge, even quantitatively, but for others we lack even a qualitative knowledge. This means, that in many cases it is not possible to relate an environmental change to a specific discharge. The implication is obvious: in these cases it is impossible to make a traditional benefit-cost analysis. If, however, the uncertainty about the interaction function and the physical damage function can be represented by a probability distribution, this distribution can be used to estimate the monetary damage.

Chapter III

ECONOMIC FUNDAMENTALS AND DEFINITIONS

A. SUMMARY

This chapter presents the fundamentals of economic theory necessary for a proper understanding of the nature of damage functions. Without such a foundation, one cannot be sure that one's estimates of damage are meaningful. The approach used is the standard neo-classical approach in economic theory. This handbook does not discuss the pros and cons of neo-classical theory, but not alternative approach to 'monetary damage functions' appears possible. Any scepticism about this approach would also imply the existence of scepticism in the use of monetised benefit-cost studies in the domain of the environemnt. Moreover, all existing studies which have used monetary damage functions are based either explicitly or implicitly on such an approach.

Several economic terms and concepts are defined within this chapter, the most important of which are presented here:

<u>The compensating variation</u> for a change in environmental quality is the least amount of money necessary to compensate an individual for the change.

<u>The total compensating variation</u> for an environmental quality change is the sum of the compensating variations for all individuals affected by the change.

<u>The monetary damage</u> is the total compensating variation of a change that is detrimental to the environment.

<u>The monetary benefit</u> is (the absolute value of) the total compensating variation for an improvement on the environment.

<u>The financial loss</u> is that part of the monetary damage which is realised by changes in expenditures.

<u>The amenity loss</u> is the difference between monetary damage and financial loss.

<u>The monetary damage resulting from disease and premature death</u> is the total compensating variation for a change in the probability distribution of contracting the disesse and dying prematurely.

B. THE MARKET MECHANISM

The basic philosophy underlying the use of markets as allocation mechanisms is a belief in consumer sovereignty. Through their purchase of commodities and services, consumers presumably have the ability to allocate resources so that production will correspond to the preferences of the consumers. The co-ordinating instruments in this process are prices. When a consumer decides how much of a given commodity he wants to buy, he first compares that commodity and its price with other commodities and their prices. The amount the consumer buys of a particular commodity is such that if he were to buy one more unit of that commodity he would receive less benefit than if he were to buy other competing commodities. If, for example, he were to buy an additional unit of meat, the increased expenditure for meat would be equal to the meat price. If the consumer does not buy the extra unit of meat, we can conclude that he prefers the extra quantity of vegetables to the extra unit of meat. The relative price of meat in terms of vegetables reflects the values the consumer puts on marginal units of meat and vegetables.

This principle can be formulated in another way. Let us define the _marginal willingness-to-pay_ for a good as the amount of money the consumer is willing to pay for one more unit of the good. The marginal willingness-to-pay for a good is then equal to the price because, if the marginal willingness-to-pay were greater than the price, the consumer would buy more of the good and vice versa.

The prices determined in markets will thus reflect the consumers' marginal willingness-to-pay for different goods. This interpretation of market prices is necessary for a proper understanding of environmental monetary damage functions. These will be defined so as to reflect this property of market price as much as possible.

We extend our basic description of consumer behaviour by defining the _marginal compensation_ for a good as the smallest sum of money necessary to compensate an individual for the loss of one unit of that good. It is clear that the marginal compensation is equal to the price, because if the marginal compenstation were less than the price, the consumer would not have bought the unit, and vice versa. The marginal compensation is thus equal to the price and also equal to the marginal willingness-to-pay.

So far we have discussed only consumers. Let us briefly turn to producers. Assuming profit maximising firms and perfect markets for the time being, the producer will supply an amount such that price equals the marginal cost of producing the commodity. If the supply were less than this quantity, it would be profitable for the firm to increase its production because revenue would exceed the costs of an increase in production. If, on the other hand, the supply were greater, it would be profitable to reduce the output. This reasoning holds true only if markets are perfectly competitive so that the individual firm has no influence on prices. This assumption may not be valid in many circumstances, and whenever the assumption is not realistic, there will be problems in the estimation of damage functions. The diccussion in the next sections in this chapter do not rest on this assumption, however, and the rigorous definition of environmental monetary damage function presented below is independent of any assumption about market structure.

It is now evident how perfect markets operate to allocate resources among competing uses. Given the prices, competing firms will use resources up to the point where the price of output is equal to the marginal cost, i.e., the cost of producing one more unit. The competitive prices will reflect not only the consumers' preferences but also the marginal costs of producing these different goods. A change in one of the determinants, in the cost of producing meat for example, will affect the prices by changing the supply of meat which will directly affect the price of meat. In the final equilibrium, the consumers will demand different quantities of the various goods, and their marginal willingness-to-pay for these goods will have changed, but the final prices will still reflect consumer preferences.

C. COMPENSATING AND EQUIVALENT VARIATION

So far, we have discussed marginal changes only and our conclusion was that the price of a good is equal to the marginal willingness-to-pay for that good, which is also equal to the marginal compensation for the good. Let us now turn to non-marginal changes.

The compensating variation is defined as the least amount of money necessary to compensate a consumer for a change from the present situation to some other situation. For example, if the new situation is characterised by a higher price of a commodity the compensating variation is the amount we have to give to the individual so that he does not feel worse off in the new situation. When the price rises, the consumer will adjust to the new prices, and he will probably buy less of the good with the higher price and more of other goods. This adjustment can be described in the following way. First, the price increase means that the real purchasing power of the individual has fallen; he is worse off. This effect is the income effect. This fall in real purchasing power implies a reduced consumption for most goods and a different composition of the bundle of goods demanded. The latter is a substitution effect. The substitution effect remains, however, so that the individual might be buying a different set of goods than he was buying in the initial situation.

If the price of a good falls, the compensating variation is defined in exactly the same way, except that it is negative. It tells us how much we can confiscate from the individual without leaving him in a worse situation than in the original situation.

The equivalent variation is defined as the amount of transfer in money income that the individual regards as equivalent to the change from the original situation to the new one. If, for example, the price of a good has increased, we could ask: by what quantity can we reduce the income of the individual instead of raising the price of the good so that the individual is indifferent to these two changes? If the price has fallen, we could similarly ask by what quantity must we increase the income of the induvidual instead of decreasing the price so that he is indifferent between the increase in income and the fall in price? The necessary changes in income in these two cases is the equivalent variation.

The difference between compensating variation and equivalent variation is thus that the former takes the initial situation as a basis for a comparison of income changes and price changes while the latter takes the final situation as a basis. Thus, the compensating variation gives the necessary money transfers to or from the individual in order to keep him as well off as in the initial situation, while the equivalent variation can be interpreted as a monetary expression of the induvidual's change in welfare.

Although, the discussion in section B was focused more or less on the concept of marginal willingness-to-pay, we have not as yet discussed the corresponding notion in connection with non-marginal changes. The reason for this is that the concept of willingness-to-pay is somewhat ambiguous. If we consider a change from a certain situation, say A, to another situation, B, the willingness to-pay for this change is, if the change is beneficial to the individual, the maximal amount he is willing-to-pay for the change. In the case the willingness-to-pay is thus equal to the compensating variation, i.e. the amount we can confiscate from the individual without leaving him worse off. If, on the other hand, we consider a resulting situation, B, which is detrimental to the individual, the willingness-to-pay is the amount the individual is willing to pay in order to avoid the change, i.e. the equivalent variation. When one uses the term willingness-to-pay, one must therefore be very careful because the meaning of the word will differ depending on whether the change is beneficial or detrimental to the individual.

The concept of compensating variation is closely related to the concept of compensated demand curves. While an ordinary demand curve gives the quantities of a good demanded at different prices when income is held constant, a compensated demand curve gives the quantities demanded at different prices when an individual is compensated so that he is always as well off as in the initial situation, i.e. the income effect is offset by compensations. The distinction between the two concepts is illustrated in Diagram 2.

E - E is the ordinary demand curve. It shows how much the individual demands at various prices. Let us assume that the present price is p'. The individual then demands the quantity q'. If the price were to increase, the consumer would reduce his demand along the curve E - E. The increase in price leads to a reduction in the well-being of the individual. If the individual were compensated for this reduction through a money transfer, he would (in most cases, at least) buy more of the good than if he were not compensated, because of his increase in income. The compensated demand curve E' - E' thus lies to the right of E - E at prices above p'. Similarly, if the price were to fall below p', the consumer would be better off. If his gain were confiscated he would demand less. At prices below p', the compensated demand curve is to the left of the ordinary demand curve.

For example, let p' be the present price of meat so that the individual buys the quantity q'. If the price were to increase to p, for example, the consumer would reduce his demand to q. He is less well off with less meat and no compensation for this reduction If the individual were compensated for this reduction in meat, he would probably buy more meat, his income being larger, and the compensated demand curve indicates that at a meat price of p he would buy quantity q" of meat. If, however, the price were to fall to p" (below p'), the consumer would ordinarily demand more meat (q"'), but if some of his gain were confiscated his increased meat demand at at the lower price would be less (q"") than if his gain were not confiscated.

If the good were not supplied at all, or what would be equivalent, if the price increase were so great as to reduce the demand to zero, the consumer would find himself at point B in the diagram. The curve E' - E', the compensated demand curve through A, shows the relation between demand and price when the individual is so compensated that he is always as well off as he was in the initial situation. This curve intersects the price-axis at C. In Appendix A it is shown that the compensating variation is equal to the area of the triangle ACD. If the individual could not buy the good because of the absence of supply or too high a price, it would be necessary to compensate him with an amount equal to the area ACD.

Where income elasticity of the demand is small, i.e. when the changes in demand due to compensations or confiscations are small, it is reasonable to assume that the compensated demand curve coincides with the ordinary demand curve. The compensating variation will then be the area under the ordinary demand curve, ABD. In most applications it is impossible to observe the compensated demand curve; only the ordinary demand curve is discernable, and thus our empirical data may force us to use the ordinary demand curve to estimate the compensating variation.

Diagram 2

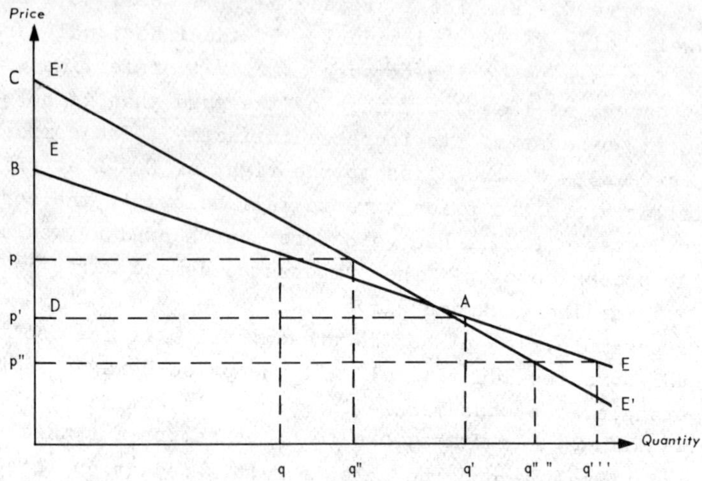

The same data problem may force us to identify compensating variation with equivalent variation. In fact, in most applied studies to date, the distinction between the two concepts is not made simply because it has been impossible to maintain it in empirical work. But even if it is impossible to distinguish between the two concepts in practice, it is important to keep them separate at a conceptual level. First, there do exist situations in which it is possible to maintain the distinction in empirical work. Second, it is often possible to know from an a priori argument which of the two concepts is greater in value and how the obtained estimate is qualitatively related to compensating variation and equivalent variation respectively. It may, for example, be possible to say that the estimate is greater than the compensating variation but smaller than the equivalent variation. In such cases one obtains bounds for the true values which may be useful. Third, in some of the literature on the subject, the two concepts are distinguished, and as an aid to the reader, it has both concepts presented in this handbook. Finally, and perhaps most importantly, there is a need for a clear and logical foundation for monetary damage functions, and in such a foundation it is necessary to distinguish between compensating variation and equivalent variation.

D. MONETARY DAMAGE FUNCTIONS - ONE INDIVIDUAL

In chapter II B, the monetary damage arising from a change in the environment is defined as the least amount necessary to compensate the individuals concerned, i.e. the compensating variation. This concept is elaborated and exemplified here.

The basic reason for choosing this definition is found in economic theory. In order to judge whether a certain change in the economic system or in the environment is beneficial or not, it is necessary to have some criterion. The present approach assumes that individuals' desires should matter, and any environmental change should be considered beneficial if it is judged to be beneficial by the individuals concerned by the change. In most cases, however, some individuals are hurt by the change while others benefit. If we restrict our discussion to those changes which benefit all individuals, we can in an affirmative way declare that the changes are beneficial. Such changes are said to satisfy the Pareto criterion. The importance of the Pareto criterion is that it permits us to make judgments about certain actions without having to specify interpersonal value judgments.

Our definition of monetary damage is chosen so that it will conform to this theoretical concept. The following example may help to clarify the definition. Let us assume that a firm is planning to extend its operations in a way that would degrade the quality of a stream. In so doing, the firm would earn an extra profit equal to $1,000. On the other hand, the reduced stream quality would cause damage to an individual living along the stream. Let us assume that the monetary damage to the individual is $700. It is then possible for the firm to compensate the individual so that he is not worse off because of the change in stream quality and it would still have an excess profit of $300, which could be used in such a way that at least one individual would be better off. The change would in this case satisfy the Pareto criterion. If, however, the compensation were not made, the change would no longer satisfy the Pareto criterion, and it would not be possible to judge whether the contemplated change is beneficial or not. In order to be able to decide whether the extension of the operations of the firm is worthwhile from a social point of view, it is necessary to introduce some explicit value judgments on the distribution of income between the individual and the owners of the firm. It is thus clear that the excess of $300 may not mean anything. If both parties involved are regarded as being equal in the sense that it is a matter of indifference as to which party received the $300 it would be possible to judge whether the change would be beneficial from information on the monetary damage and the extra profit arising from the extended operations of the firm. This is due to the fact that it no longer matters to whom the benefits accrue; it is the total monetary value to both parties which is important. If this were positive, the project would be beneficial, and if it were negative, there would be more resulting damage than benefit, and the project should not be undertaken. These distributional effects will be further discussed in section F.

E. EXAMPLE

As a further example of the above definitions, we might consider an individual who spends his wage income on consumption. When the air quality is high, the individual is healthy. He can supply labour. He does not need medical care. When the air becomes polluted, the individual becomes sick and can no longer work. He must also spend a certain sum on medical care. What is his monetary damage as a result of the air pollution?

As indicated above, it is in general very difficult to make an empirical distinction between compensated and non-compensated demand functions. It is assumed here that it is possible to identify the non-compensated demand and supply functions(1) for medical care and labour with the compensated functions. On the other hand, we do not assume that the non-compensated demand function for consumption goods is identical to the compensated demand function. The distinction between these two functions is in a sense the key to the following discussion.

Let us denote the price on consumption goods p, the wage rate w, the price of medical care s and the environmental quality Y, and then assume that the supply function for labour can be written:

$$L(p, w, s, Y).$$

If there is no air pollution, say $Y = Y'$, the individual will supply a positive amount of labour, and his wage income is:

$$wL(p, w, s, Y').$$

This wage income is spent on consumption goods, the demand function of which is:

$$C(p, w, s, Y),$$

1) A compensated supply function for labour is a function relating the supply of labour to different wage rates when the individual is compensated for the variation in the wage rate.

and on medical care, the demand function of which is:

$$M(p, w, s, Y).$$

We assume that when there is no air pollution, the individual does not demand any medical care, i.e.

$$M(p, w, s, Y') = 0$$

irrespective of the price of medical care s. We then have the following budget restriction in case Y = Y'.

$$I + wL = pC,$$

where I is the individual's income from sources other than labour.

The question to be answered is: how much must the income I, increase in order to compensate the individual for a degradation of the air quality to Y", an air quality so deteriorated that the individual can no longer work and must demand a particular amount of medical care? Let the necessary compensation be CV (compensating variation). Let the compensated demand function for consumption goods be C(p, w, s, Y). The new budget restriction can be written:

$$I + CV = pC + sM$$

Comparing the two budget equations we find that:

$$CV = wL(p, w, s, Y') + sM(p, w, s, Y'') + p[C(p, w, s, Y'') - C(p, w, s, Y)]$$

The interpretation of this expression is straightforward. The first term on the right-hand side is the loss in wage income due to air pollution. The second term is the increase in expenditure on medical care. These two terms cover the financial loss which the individual will experience as a result of air pollution. The last term shows the increase in expenditure on consumption goods that is necessary to compensate him for the reduction in well-being which the illness has caused. We will call this amount the 'psychic cost'. The total monetary damage is thus equal to the sum of the financial loss and the psychic cost.

In this example, we have introduced the term 'financial loss'. This will be often used in later chapters. Financial losses are direct effects on income and expenditure, resulting from environmental changes. The residual monetary damage that is not reflected in the financial losses will be called psychic costs or amenity losses. The damage resulting from reduced recreational possibilities and the destruction of beautiful scenery are examples of amenity losses.

F. DISTRIBUTIONAL ASPECTS

Let us assume that there are many individuals affected by a change in environmental quality. Some individuals may regard the change as beneficial while others regard it as detrimental. If the change occurs because of pollution, the polluter most likely regards the change as beneficial (otherwise he would not have polluted the environment), and the victims regard the change as detrimental. For each individual there is a corresponding compensating variation, which is negative for the polluter and positive for the victims individually. If we sum up the compensating variations for all victims, we obtain an amount that is sufficient to compensate all of them for the environmental change. When lump sum transfers are feasible, this sum is obviously the proper measure of the monetary damage. If for example, the sum is less than the gain made by the polluter, it would be possible to compensate all of the victims completely, and there would remain a surplus which could be distributed in one way or another. The Pareto criterion would thus be satisfied and the pollution, after the compensations had been made, would be beneficial.

Let us now consider the reverse case, where pollution abatement is contemplated. For each of those who have suffered as a result of the pollution, the compensating variation is the amount it is possible to confiscate (i.e. the willingness-to-pay) from an individual without leaving him or her in a worse position. In this case we will use the term <u>monetary benefit</u> to describe the compensating variation. If the sum of all the individual monetary benefits exceeds the cost of the pollution abatement, it is possible to confiscate these benefits in order to cover the abatement cost and end up with a surplus. The pollution abatement should thus be undertaken according to the Pareto criterion. If the sum of the individual monetary benefits is smaller than the cost of pollution control, the monetary benefits from environmental improvement are not sufficient to motivate the undertaking of pollution abatement.

"If lump sum transfers are feasible, the sum of compensating variations for the individuals affected by the change in environmental quality, henceforth called the total compensating variation, is the proper definition of monetary damage". The total compensating variation, as a function of ambient environmental quality, is by definition the "<u>monetary damage function</u>".

If lump sum transfers are not feasible, the compensations and confiscations cannot be carried out and the above analysis is no longer relevant. In order to proceed, we will distinguish between two cases; first when equity questions do not matter, and second, when they do matter.

If equity is not of prime importance, an increase in nominal income is regarded as the same irrespective of the way in which it is distributed. An additional $1,000 to one person means the same to the decision-maker as an additional $1,000 to another. This assumption about the underlying values of the decision-maker can be interpreted in at least two ways. The first interpretation is that the income distribution is already optimal so that all individuals can be judged equally. The second interpretation is that there is some division of work within the decision-making organisation, so that the department which handles environmental policy need not care about income distribution. The second interpretation is more realistic. General income distributional problems are treated by general policy instruments, and environmental problems appear to be discussed with not explicit reference to equity problems.

If the distributional aspects do not matter, it is clear that the total compensating variation is still the proper definition of monetary damage and benefit. Only the total gains and losses are of interest, and these are given by the total compensating variation.

If equity problems are of concern, it is no longer possible to define the concept of monetary damage in an unambiguous way. What we would like to do is to compute a weighted total of individual welfare changes. We know from section C that one measure of these welfare changes is the equivalent variation. The equivalent variation gives the change in nominal income which is equivalent to an environmental change. If we could give to each individual a weight which corresponds to the interpersonal value judgments of the decision-maker, it would be possible to compute the total welfare change as the weighted total of the individual equivalent variations. It is thus possible to develop a formula which can be used to make consistent decisions on environment and income distribution.

For the remainder of the monograph we use the total compensating variation as the definition of monetary damage, but if it is possible to identify the monetary damage suffered by different groups, it is also possible to use the methods presented below to estimate the damage to each group. These estimates of damage can then be weighted and summed up to obtain an estimate of societal damage adjusted for income distribution. A further point should be made, however. In section C, we noted that it is very difficult in practice to distinguish between compensating and ordinary demand curves. This implies that, empirically speaking, the distinction between compensating variation and equivalent variation cannot often be made. For this reason one can roughly examine distributional aspects by weighting the monetary damage, as defined above, of different groups of individuals differently.

Although we will not discuss distributional matters any further, except for some randomly scattered remarks, this does not mean that we believe that they are unimportant. By following the outline suggested above, it is possible to treat distributional problems in a satisfactory way through use of the methods presented in chapters V and VI.

G. RISK AND UNCERTAINTY

In chapter II it was concluded that environmental effects were not always deterministic. In fact, in many cases, it is necessary to take explicitly into account the fact that the exact outcome of an invironmental decision cannot be predicted. A water pollution abatement measure which reduces the discharge of pollution in terms of BOD (biological oxygen demand) will have effects on the oxygen content in the water. However, as the oxygen content is affected by a number of other factors as well, many of which are of a random nature, the result of the abatement measure will be a new probability distribution of the oxygen content. When deciding whether the measure should be taken or not, one must decide between different probability distributions of the oxygen content. Another example is furnished by health effects from air pollution. It is not possible to predict the effect on a particular individual of an increase in air pollution but it may be possible to estimate the change in the probability of contracting a specific disease due to the increased air pollution. When considering whether the increase in air pollution is acceptable or not, it is necessary to compare different probability distributions.

In some cases, our ignorance is so great that predictions of the probability distribution of the environmental effects from certain actions are highly speculative. For such problems a monetary damage function approach does not seem to be at all fruitful and completely different methods must be used. For that reason, we will not consider this problem further.

The types of problems which we will study and which include a choice between different probability distributions are those which are usually referred to as problems with *risk*. Let us assume that we are considering a certain measure which would change the probability distribution of the oxygen content in a body of water. Let us assume also that it is possible to attach a monetary evaluation to each possible outcome, i.e. to each conceivable level of oxygen concentration. Information on the probability distribution of the outcome enables us to compute the mathematical expectation of the monetary damage. What is the meaning of such a figure? If it could be accepted as the monetary damage of the action taken, it would imply that the individual is indifferent to either a certain outcome - equal to the expected monetary damage, or an uncertain outcome - represented by the probability distribution. There are no obvious reasons, however, why the individual should be indifferent: in fact, most of the literature on the problems of risk is based on the assumption that individuals are risk averters. *Risk aversion* is best explained by an example such as the following. In a lottery in which the prizes are different sums of money, the individual may prefer a certain outcome which is equal to the mathematical expectation rather than to participate in the lottery. He is then showing risk aversion. This means that the mere existence of risk is adverse to the preferences of the individual. He would be satisfied with less than the expected value of the lottery if he could obtain that without risk. Insurance supplies another example which illustrates that many, if not most, people are risk-averters. In real terms, the total amount paid to insurance companies is greater than the expected value of the losses insured, as insurance companies must pay for overheads and make profits. The difference in real terms between the expected losses and amounts paid can be considered the risk premium.

Applied to the monetary damage resulting from a change in the probability distribution of environmental effects, this means that in general it is not possible to equate the expected monetary damage with the proper monetary damage, but we have to use the individual's evaluation of the new probability distribution to determine the proper monetary damage. Some of the technical problems involved are discussed in the appendix to this chapter.

Our basic definition of monetary damage stands despite difficulties which arise with respect to risk. The monetary damage (or benefit) from a change in the environment is the least amount necessary to compensate the individuals concerned for the change in probability distribution with regard to environmental quality.

Let us apply this approach to a particular case, Section D treated an example of monetary damage in the case of a disease contracted because of increased air pollution. The monetary damage to an individual who contracted the particular disease was composed of three parts: the wages foregone, the increase in expenditures on medical care and the psychic cost. The total damage was the amount of compensation necessary in order not to place the individual in a worse position because of the illness. If we assume that the loss in wages is $500, the increase in expenditure on medical care $400 and the psychic cost $100, the total damage is accordingly $1,000.

Let us now assume that in situation A there is no air pollution and that the probability of contracting the particular disease is zero. In situation B this probability has increased to 0.1 per cent becasue of increased air pollution. The expected monetary damage in situation B is thus:

.9 times zero + .1 times $1,000 which equals $100.

If the individual is a risk averter his damage is greater than $100. To this amount we must add a risk premium which covers the increase in risk to which the individual is exposed. It is not possible to compute this risk premium from the information given so far, but let us assume that the premium is $50 for this example. The proper monetary damage in this case is thus $150, which is the amount with which we must compensate the individual in order to make him indifferent to situations A and B.

There is one aspect of this subject which may give rise to difficulty. Let us assume that there is a non-zero probability of premature death due to the disease. Supposing that the conditional probability of dying from the disease is 0.01, the probability of premature death is accordingly 0.1 times 0.01 or 0.001. Is it possible to compensate an individual now in order to make him indifferent to a certain death one year from now?. Common sense would say no, and it is obvious that the question itself if almost meaningless. The proper question is: how much is needed to compensate an individual for an increase in the propability of dying? We know that people in their every-day life make decisions which weigh probabilities of being killed. Each time an individual decides to travel in a motor vehicle he knows that there is a non-zero probability of a mortal accident, but the benefits from travelling in a vehicle outweigh this risk. Some part of the earnings of a taxi driver may be regarded as a compensation for the risk he is willing to take. It is clear that such compensations do in fact exist. Because it is impossible to eliminate all risks, People must make comparisons between the benefits and the degree of probability of being killed, in order to choose a particular alternative. We will therefore accept the possibility of attaching a monetary value to a probability distribution, even if the distribution were to contain a positive probability for premature death.

In the appendix to the chapter on health effects and premature death, we give a detailed analysis of the case where an increase in air pollution leads to an increase in the probability of contracting a respiratory disease which in turn could lead to death. There it is concluded that the monetary damage of this change in probability is composed of three terms:

a) the expected total financial cost associated with the disease, i.e. productivity loss and expenditure on medical care;
b) the expected total financial costs associated with premature death; this consists principally of the taxes the individual would have payed if he had survived and his net wealth accumulation if he had survived;
c) the psychic cost, which is a premium above the financial costs and is necessary to compensate the individual for the increase in risk.

In addition to these, there are also the costs suffered by the near relatives and friends of the affected person.

H. INTERTEMPORAL CONSIDERATIONS

So far, no explicit mention of intertemporal factors has been made. To a large extent the definitions given in the earlier sections in this chapter can be carried over to the intertemporal context, but intertemporal considerations introduce an additional problem to which the solution is not clear.

It is the present value of future damage which is relevant in any decision-making problem. The choice of an interest rate faces the same problems as does any other public investment decision, and requires no further analysis in this study. For most cases involving reversible environmental effects and durable equipment for pollution abatement, the standard theory for investment decisions can be applied.

There is, however, one specific problem which arises in environmental protection decisions, particularly with respect to irreversible changes in the environment. These make it necessary to look into the very distant future. In principle, the time horizon is infinite. The resulting empirical problems are obvious. How do we estimate the current value of damage occuring in the distant future as a result of actions taken today? This damage will in general depend on the values of future generations, about which we know little.

The usual procedure in cost benefit analysis by which uncertain variables are replaced by their expectations will obviously yield biased results when the individuals concerned are risk averters. But even if all were risk neutral, this procedure may yield biased results when project includes irreversible changes. It has been shown by Henry and by Arrow and Fisher(1) that replacing a random variable with its expectation will in this case very likely imply a bias in favour of accepting the irreversible change. With the definition of monetary damage given above, which involves a lump sum compensation for all consequences in the present as well as in the future, this problem will not show up. In practice, however, one has to estimate the costs and benefits for each time period which implies that the standard technique is likely to yield results biased against preservation. Therefore, whenever irreversible changes are involved, and uncertainty prevails, the whole stochastic

1) Henry C. (1974) Investments Decisions under Uncertainty: The "Irreversibility Effect", The American Economic Review, December 1974 pp 1006-1012. Arrow, K.J., and Fisher, A.C. (1974) Environmental Preservation, Uncertainty, and Irreversibility, Quarterly Journal of Economics, May 1974, pp 312-19.

nature of the decision problem should be taken into account instead of just replacing random variables with their expectations.

This consideration is valid whether or not monetary damage functions are used as a tool in environmental policy. Whenever a decision which involves irreversible effects which affect future generations is made, the decision must be based on the beliefs and values of the present generation. But there is one fundamental difference between the monetary damage function approach and a direct policy decision. The situation is in many ways similar to the problem of income distribution associated with the present generation. When we discussed that problem in section F, we concluded that it is possible to treat distributional aspects if there exists a set of weights, which if applied to the damage of individuals, will allow the summation of individual damage into a welfare measure of damage. The same approach can be applied to intertemporal analysis. We need a set of weights, through use of which we can evaluate the total welfare implication of a change in the environment. This evaluation would consider the effects on both the present generation and the future generations. In principle, at least, it would not be impossible to obtain a set of weights from the decision-maker. He should provide them for the intertemporal situation in the same manner as for income distribution considerations. In fact, the interest rate can be considered as one set of weights or discount factors for the consideration on intertemporal effects. In contrast to the temporal case, it is not possible to estimate the damage for individual future years, to which the weights should be applied. Finally there are no methods which can be used to reveal the preferences of future generations. The use of monetary damage estimates, which reflect the preferences of the present generation, to make decisions on irreversible environmental changes means that we assume that preferences will remain constant among cohorts and that the preferences of future generations should have less weight than those of the present generation.

Furthermore, a market interest rate reflects the trading possibilities between different individuals in the present generation. Such trading is not possible between people living now and in the future. This remark is of special importance in connection with irreversible changes in the environment. Future generations can not buy back a certain piece of nature which has been permanently developed for industrial purposes. The use of a market interest rate in evaluating the monetary damage may therefore not be very meaningful.

Chapter IV

CLASSIFICATION OF DAMAGE

A. INTRODUCTION

Damage can be classified according to the receptor suffering the damage, the pollutant causing the damage, or the sources emitting the pollutant. In the present study the focus is on the receptor and how damage to it can be related to environmental changes, and ultimately to the sources of these changes. Accordingly, damage is classified here into six broad receptor categories, and the major sub-categories of damage are given for each major category. Within each category, two general types of damage are considered, financial losses and losses of amenity. The definitions of these categories correspond to the definitions given in Chapter III. It should be noted that in reality the dichotomy between these categories is not always a sharp one, and in some cases it is impossible to obtain separate estimates of the two components. (See Chapters V. and VI.) To date, studies have used other terms to denote this dichotomy as "direct and indirect costs", "economic and social costs", "tangible and intangible costs", and "physical and psychic costs".

It should be noted that classification schemes other than that presented here may be equally acceptable. The final administrative use of the damage information will dictate the optimum classification scheme.

Some notion of what should be considered in a specific damage category is indispensable to the task of evaluating an estimation procedure or an estimate itself. The present choice of classification scheme indicates the types of damage which must be considered in any comprehensive study of environmental damage functions. It also provides a set of reference points for the various studies and methods considered in the next chapter.

B. HUMAN HEALTH

The establishment of a causal relationship between environmental pollution and health effects is difficult, and no attempt will be made to solve the problem or to suggest criteria which might be used to overcome it. Some estimate of this relationship is nevertheless necessary before the following monetary losses are calculated.

1. Financial losses

The major financial losses of pollution-induced health effects are due to productivity losses and increased costs of health care.

i) Productivity losses result from accelerated mortality rate as well as from pollution-induced morbidity and include reduced productivity of an affected person who nevertheless continues his usual daily activities. The lost productivity in services not usually included in GNP is also important. These include such activities as those performed by housewives and students.

ii) Increased costs of health care should also include social overhead capital costs and some of the costs of the research undertaken to establish the etiology and optimal treatment of the resulting health effects.

iii) Other financial losses due to the increased probability of ill health and death are relatively minor compared with the above two. However, two which have received some attention are moving costs incurred from migration by those who wish to avoid adverse health effects, and the cost of premature burial due to pollution-induced premature death.

2. Losses of amenity

Amenity losses are the incremental costs over the financial losses resulting from the increased probability of ill health and death, as a result of environmental quality changes. These losses are the costs of suffering, bereavement and limitation imposed upon the affected individuals, their families and their society. The definition of ill health here is a broad one and includes annoyance from a nuisance such as noise or odours.

C. FAUNA

The damage considered in this category represents financial and amenity losses incurred as a result of pollution-induced effects on non-human animal life; these effects are, however, valued by human beings.

1. Financial losses

The principal financial losses are the value of lost production due to reduced animal or fish population and/or to the reduced size of certain species. The effects may be indirect. For example, the pollution induced deaths of birds may allow increased insect populations which in turn cause increased agricultural damage.

2. Losses of amenity

These centre principally upon the losses in amenity from decreasing (and even from the extinction of) bird and animal populations and from decreased pleasure from fishing and hunting because of reduced or protected populations.

D. FLORA

1. Financial losses

Lost or reduced production from pollution-induced effects on agricultural, horticultural and forestry products is the predominant source of damage in this category.

2. Losses of amenity

The disamenities of horticultural damage and diseased trees are the principal social costs in this category.

E. NATURAL RESOURCES

Damage in this category excludes the effects on life and restricts itself to effects on the natural (as opposed to manmade) physical resources. Air is not included among the resources considered.

1. Financial losses

Lost production due to polluted soil or water is one of the economic costs incurred.(1)

2. Losses of amenity

The losses of amenity are principally due to decreased recreation and aesthetic benefits. In some cases a loss in option value may also contribute significantly to this category.

1) To allow a consistent, mutually-exclusive classification scheme, damage is classified under the first receptor in any chain of damage.

F. MATERIALS

Pollution has several types of effects on materials: it can corrode them, deteriorate them, change their colour, soil them or necessitate special measures to protect them.

1. Financial costs

There are several types of costs which can be classified here, the most important of which are a result of:

i) the reduced longevity of a material (material and installation costs are both considered here);

ii) the decreased productivity or utility of a material;

iii) the necessity to design a satisfactory substitute material;

iv) losses due to an inferior substitute;

v) the need to protect the vulnerable material from sources of pollution;

vi) any additional maintenance, including the cleaning of a material, required because of pollution.

2. Losses of amenity

The major losses of amenity from pollution effects on materials are those of enduring and suffering soiled, damaged or inferior products and materials because of pollution. It should be noted that these costs can be very high for certain selected monuments and objects to which society attaches high aesthetic value.

G. CLIMATE AND WEATHER

This category is included principally for completeness. Discussion of pollution effects on climate and weather is largely speculative, although there is some evidence of effects on rainfall due to air pollution and smoke and London has experienced a remarkable increase in sunshine as its levels of smoke pollution have decreased. There is also concern that the increased release of CO_2 and particulates into the atmosphere could raise or lower the global temperature respectively. Any significant change in the global temperature could, of course, create disturbances and damage. Given the present high level of uncertainty attached to these effects, they will not be considered here.

1. Financial costs

Increased or decreased rainfall could affect agricultural output and some outdoor production work, both of which could have important financial effects. Decreased sunshine could lead to increased expenditure for artificial lighting. Reduced visibility from air pollution also contributes to automobile (or even aircraft) accidents and disruptions in road and air traffic, both of which result in financial losses.

2. Losses of amenity

Increased rain or decreased sunshine would certainly lead to losses in amenity through decreased recreation opportunities, for example. Reduced visibility from pollution would also contribute towards losses in amenity where visibility loss is responsible for accidents, traffic disruptions or decreased pleasure in landscape viewing.

H. OTHER CONSIDERATIONS

Within each of the above categories, risk aversion and loss of option value are potential contributors to amenity loss and should not be ignored. They can be significant factors in amenity loss.

Table 1 summarises the types of damage considered in the classification scheme presented.

Table 1

Damage Classification scheme with examples

Damage Category

	Human health	Fauna	Flora	Natural resources	Materials	Climate and weather
Financial loss	Productivity losses; health care costs including increased research costs to avoid pollution; premature burial costs	Lost animal and fish production	Reduced crop production; reduced forest growth	Lost production from polluted water or soil	Reduced life of a material; reduced utility of a material; cost of producing a material; extra cost of a substitute	Reduced agricultural output from decreased rainfall; larger lighting expenses from decreased sunshine
Loss of amenity	Risk aversion cost of suffering; cost of bereavement; cost of limitations imposed upon an individual, his family and his society	Risk aversion reduced pleasure from fishing and hunting; reduced wildlife populations	Risk aversion reduced pleasure from horticultural and forestry losses	Risk aversion decreased recreation benefits	Risk aversion endurance of soiled or damaged materials; damage to selected aesthetic monuments and objects	Risk aversion decreased recreation benefits from decreased sunshine or increased rainfall; decreased pleasure from reduced visibility

Chapter V

ESTIMATION OF FINANCIAL LOSSES

A. SUMMARY

Chapters V and VI form the bulk of the monograph. They discuss methods which are used to apply monetary values to physical damage. The methods are discussed both in terms of the underlying economic theory and in terms of their application. The principal objective of the theoretical discussion is to relate the estimates produced by the various methods to the definitions given in chapter III. It will then be possible to evaluate the various methods in terms of how well they approximate the theoretical objectives in the estimation of damage.

The wealth of existing studies serves as a useful background for the analysis of the various methods. First of all, existing studies serve to illustrate the methods presented in this chapter. The examination of existing studies can also suggest the practical difficulties involved in employing given methods, thus allowing an evaluation of various methods in the light of existing experience. Finally, in some cases existing studies indicate ways in which various problems can be overcome.

No one method is superior to all others in evaluating monetary damage. In estimating the damage from a specific change in environmental quality the best approach to use is probably to combine the use of several methods. We know from chapters III and IV that damage can be divided into several components. In the example analysed in chapter III A. the monetary damage from the contraction of a disease could be broken into three components: wages foregone, expenditures on medical care and the loss of amenities. No one method estimates each of these terms best, hence a method must be used to estimate each component. Because the existing methods are often relatively crude, there can be a need to compare the results obtained through one method with results obtained through others.

In this chapter we concentrate on direct methods which are generally used to estimate financial losses. The following chapter presents methods which are generally applied to estimate amenity losses. In spite of this division the dichotomy between applications of methods is not a sharp one. Some methods described in the next chapter, e.g. questionnaires, can be used to estimate financial losses. In some cases direct methods may yield estimates of amenity losses. The division remains however, an expedient one, as the approaches for estimating financial losses and losses of amenity differ for the most part.

B. THE ESTIMATION OF FINANCIAL LOSSES: GENERAL CONSIDERATIONS

Financial losses were defined in chapter III as the monetary value of changes in the demand for marketable goods and services due to changes in the environment. We have already seen examples of such losses, e.g. wages foregone, expenditures on medical care, productivity losses.

There are four problems related to this approach, namely:

i) the specification of effects, or the problem of identifying marketable goods and services which are affected by a change in the environment;

ii) the relating of effects to a specific level of environmental deterioration;

iii) the problem of finding the proper prices, including the interest rate, which should be used in a monetary evaluation;

iv) the problem of interpreting the calculated financial values and relating them to the monetary damage in the context of the problem under study.

1. Specification of effects and relating them to environmental quality

The first two problems are to some extent the same as finding the physical damage function. We must know the physical effects of the change in the environment in order to be able to say something about the effects on the demand for marketable goods and services. Furthermore, the physical damage function may not sufficiently specify all effects. Physical effects can induce further behavioural changes or effects which should be specified. Increased air pollution may, for example, lead to more use of air conditioning. This is not customarily included among the physical effects. We must know both the physical damage function and the induced effects on individual behaviour before we can evaluate the financial losses.

In general, the handbook does not concern itself with the first two steps. It considers them as inputs in constructing monetary estimates. However, the constructor of monetary damage functions does have some responsibility for the execution of this work. He must review the inputs and determine whether there are any major omissions in the effects specified. He must know how to relate the information he has on environmental quality to the objectives of his work.

A brief consideration of the first two problems may provide some appreciation of the problems confronted in realising them. The constructor of monetary damage functions may then realise that his task is no more difficult than that of completing the first two steps.

The first step in estimating any losses is the specification of the physical effects, the induced effects in human behaviour and losses in amenity. In this chapter we will concentrate on the first two types of effects. In most cases, there is little or no agreement on the effects which do occur, and the information which does exist is of a qualitative rather than a quantitative nature. A rapid glance at existing studies indicates the range of disagreement on environmental effects. One important study of air pollution considers only two health effects, lung cancer and bronchitis, and for each of these categories there exists a divergence in the expert opinion which contributed to the study. For example, in the case of bronchitis linked to urban air pollution, some experts felt that up to one-third of all urban cases are due to air pollution, whereas others felt that there were no cases of bronchitis due to air pollution. Other studies of environmental health damage attribute many more health effects to air pollution, adding such categories as arteriosclerotic heart disease, the common cold, asthma, emphysema and even deaths due to motor vehicle accidents. Moreover, the definitions of financial loss and loss of amenity differ between studies. With respect to health losses, one study believes costs of premature burial to be sufficiently important to be included among estimated financial losses. Othor studies have not generally considered these costs. Also in the health area, several studies have ignored amenity losses altogether.

The second step, that of relating effects to a specific level of environmental deterioration, is not any simpler than the first step. It is difficult to describe environmental quality in a succinct manner for a given time, space and pollutant distribution - even if this distribution were known. The resulting description depends greatly upon the parameters and sampling points selected and upon the method chosen to summarise these inputs. It is often unclear whether long-term averages or acute episodes relate better to an effect. There is often disagreement on measurement techniques or on where to measure a pollutant. The resultant measure may or may not be a good approximation of the true exposure. In the area of health and air pollution we would like to monitor each individual's exposure, indoor and outdoor, and relate this to the effects he experiences. As this is impossible some measure of outdoor environmental quality which is thought to be representative of population exposure is chosen. There are problems involved also in attributing effects solely to one pollutant. Systems of pollutants, pollutant-meteorological or geological combinations and pollution-behaviour interactions may be very important in determining environ ental damage. For example, precipitation through increasing runoff may aggravate the eutrophication potential of a lake; or smoking can alter one's risk of contracting a respiratory disease in the presence of pollution.

Economists have usually grappled with these problems through the use of an input from an expert who has, to the best of his ability, provided a list of effects for a given level of environmental quality. In some cases economists have surveyed expert opinion in this matter and tried to reach a consensus. In at least one study a poll of experts was used to construct a probability distribution of effects which in turn led to damage estimates which were accompanied by measures of their variability.

The problem of describing environmental quality is usually overcome by simply defining it to be the existing environmental quality of a certain period and place. This description may be useful for the particular study undertaken but it makes it difficult to transfer results to other situations.

2. The choice of a proper price and the interpretation of estimated values

As financial losses are by definition related to changes in the demand for marketable goods and services, prices can be used in order to assign a monetary value to these changes. In principle, there are no problems present here, although the empirical problems involved in finding the correct price may sometimes be quite substantial. When the partial equilibrium assumption is not valid and prices change because of a change in the environment, severe empirical problems can arise. In order to calculate the financial loss in this case, it is necessary to estimate the demand functions if possible so that the compensating variation can be computed as the area under the demand curve. In chapter III it was shown that the compensating variation for a change in one price could be presented by the area A C D in Diagram 2.

Other problems may arise in the choice of a price when prices are not the equilibrium prices in competitive markets and when indirect and direct taxes distort the price relations. In the first case, the prices should be corrected if possible. For example, if an agricultural policy maintained agricultural prices at a higher level than they are at the equilibrium, the existing prices should be replaced by estimates of the equilibrium prices. All prices should be net of taxes except in those cases where taxes are used as corrective instruments for external dis-economies.

When the partial equilibrium assumption holds the calculation of financial losses is relatively straightforward: the total losses are multiplied by the price per unit of loss. Let us consider the cases where crop losses were estimated in two states, California and Pennsylvania, in the Unites States. In these cases trained crop surveyors examined the crop damage and gave estimates of the losses due to air pollution. These losses were multiplied by the average seasonal price of the given crops, unprocessed, at the first delivery point. Hence, on the assumption that the magnitude of the loss is not sufficiently large to change the price of the agricultural product, the financial losses are equal to the revenue lost to the farmer. The assumption appears reasonable here since the losses represent a small fraction of the value of the crops. But if the loss for a specific crop were to diminish the supply of that crop significantly, then the reduced supply could lead to a higher price and the application of average unit prices to the lost crop would overestimate damage costs. A lower alternative price which reflects the equilibrium price then needs to be substituted for the average seasonal price.

If there were an agricultural policy which imposed artificial prices upon these crops, then the existing artificial price should not be applied, but some estimate of the real price.

The problem of interpreting the estimated financial loss is difficult because financial losses are in many cases only part of the monetary damage, amenity losses representing the other part. Thus the relationship between financial losses and monetary damage is not a simple one-to-one relationship and cannot be discussed in general terms. For this reason there is a systematic discussion of some of the possible financial losses below. Nevertheless, in most cases it is necessary to supplement the calculated financial losses with estimates of amenity losses.

C. PRODUCTIVITY LOSSES

1. General approach

These losses are associated either with materials damage (including effects in agriculture and forestry) or with health effects. As the latter effects will be discussed in another context, we will consider materials damage here. Materials damage may cause a reduction in the rate of production, in which case the value of this loss in production may be taken as a measure of the financial loss if no offsetting measures are taken. If the materials can be protected by painting or coating them, then it is possible to use the cost of protecting them as a measure of the financial loss. If the damaged materials can be replaced by a substitute that does not become damaged, then the cost of such substitution may be taken as a measure of the financial loss.

We know from chapter III that if the productivity loss does not induce a price change, the proper monetary damage is given by the reduction in the profits of the firms concerned. In most cases the measures of financial loss will be good approximations of these profit reductions, the exceptions occurring when the firm adapts many of its decision variables to materials damage. If, for example, the firm decides to replace materials, partly because they suffer from environmental effects and partly because it wishes to use a newer substitute which will increase productivity it would be incorrect to estimate the monetary damage to the replaced materials as the costs of replacing them. This would lead to an overestimate of materials damage.

The following diagram presents the marginal cost curves before and after materials damage has occurred. If the amount of production $\bar{x}$ is unaffected, the financial loss is equal to the area ABCDEF, which corresponds to the costs of replacing the damaged materials. It is, however, optimal for the firm to reduce its production to x, and the monetary damage is equal to the area ABDEF. The financial loss, measured as the cost of replacing the damaged materials, provides an overestimate of the monetary damage.

If several plants within an industry suffer materials damage and if they all react by changing their production volumes, this in general will affect the production price and the loss in profits or producers' surplus is no longer sufficient as a measure of the monetary damage. It is also necessary to take the loss in consumer surplus into account. The total compensation necessary for the air pollution damage consists thus of two parts, first a compensation to the producers for the reduction of their profits and second, a compensation to the consumers for the resulting price change. It is obvious that in many cases the calculation of this total compensation will be almost impossible.

In some cases the effect on producer surplus will be capitalised into land values. In this case, the profit will be unchanged and the total monetary damage is equal to the decrease in land values. This matter will be discussed in the next chapter. However, it seems safe to assume that changes in the volume of production and similar adjustments can be neglected and the direct financial loss, measured as above, is a proper approximation of the monetary damage. Air pollution damage to materials is concentrated close to major

Diagram 3

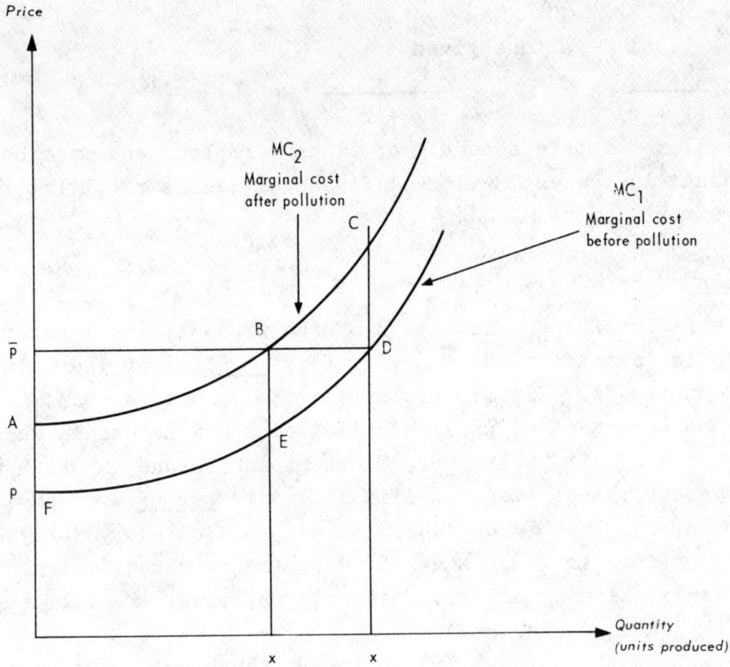

sources of the air pollution: since industry is widely dispersed, much of it is unlikely to be in the high pollution area and it seems reasonable to assume that product prices will seldom be affected by materials damage.

It is important to be aware of the difficulties which exist in measuring the financial loss; for example, how can the cost of replacing damaged materials with substitutes be calculated? The expenditure for new materials is an appropriate measure of damage only if the existing materials would have had an infinite life in the absence of pollution. If, however, the lifetime of a material in a clean environment is finite, the above procedure must be adjusted. Assume that the remaining lifetime of a material is ten years in the absence of pollution, but that pollution has necessitated its replacement after two years (instead of ten) by a new substitute material which costs 100 units more, the present value of the replacement cost is:

$$\frac{a + 100}{(1 + r)^2}$$

where a is the cost of the old material and r is the interest rate. We assume that the new material is not affected by pollution so that we need not consider future replacements due to pollution. If the environment were not polluted the present value of the replacement cost would have been:

$$\frac{a}{(1 + r)^{10}}$$

The damage cost C is thus given by:

$$C = \frac{a + 100}{(1 + r)^2} - \frac{a}{(1 + r)^{10}} \qquad (5 - 1)$$

In general, the future schedule of planned replacement must be taken into account in the estimation of financial losses resulting from premature replacement.

2. Example

An example of the calculation of the productivity loss of materials is taken from the study by Fink *et al*(1) of the Battelle Memorial Institute. Expert opinion and the results of controlled experiments were used to provide a list of those materials which were affected by air pollution. The approach is one in which component systems rather than specific materials *per se* are considered. For example, instead of considering steel as a category in itself, component systems in which steel is used are examined: steel storage tanks, highway and railroad bridges, power transformers, chain-link fencing, etc.

The vulnerable systems are defined through an extensive screening procedure. The two broadest categories, Construction, Division C and Manufacturing, Division D, of the Standard Industrial Classification (S.I.C.)(2) are identified as those in which all relevant component systems containing metal would occur. Experts then eliminate those specific component systems which are relatively rare, unexposed to air pollution. A comprehensive screening procedure narrows this extensive list down to nine major components. Other metals exposed to air pollution are considered economically unimportant or relatively insensitive to air pollution.

The losses due to air pollution are of two types: (1) the additional protection and maintenance expense required to protect exposed systems from air pollution, and (2) the costs of the decreased life of the system resulting from air pollution-induced corrosion. The losses of the first type are much larger because if air pollution significantly increases the rate of corrosion, then maintenance and protection measures are usually taken to overcome this problem.

---

1) Fink, F.W., F.H. Buttner and W.K. Boyd (1971) "Final Report on Technical Economic Evaluation of Air Pollution Corrosion Costs on Metals in the U.S.", made to the Air Pollution Control Office, Environmental Protection Agency, Battelle Memorial Institute, Colombus, Ohio.

2) This classification scheme is similar to the I.S.I.C. (International Standard Industrial Classification) system.

The marginal maintenance costs are essentially those additional painting costs necessitated by the air pollution. To derive these costs in units of total weight shipped, the I.S.I.C. data are converted into equivalent exposed areas by engineering methods which depend upon the component considered. As an example, let us consider the calculation of increased maintenance costs of steel storage tanks. Different categories of tanks such as tanks for water, for petroleum and for chemical storage are considered separately. For these categories estimates are made of the total surface area of prototype (typical storage tanks) on the basis of I.S.I.C. data. Estimates are then made of the percentage of the total surface area exposed to pollution and the resulting area is then multiplied by the annual extra cost of protection by paint per unit area, to obtain the total natural loss in that category. The annual extra cost per unit area is calculated by determining the typical useful life of the structure considered from an appropriate Internal Revenue Service Table. The painting costs of maintaining this structure throughout its lifetime are calculated for rural and for polluted conditions and the difference in the two costs is converted to an annual cost by dividing by the normal life of the structure. In the case of water storage tanks the expected life is fifty years and a coat of paint lasts about twelve years in a rural environment and eight years in a polluted environment (with an average level of 0.03 - 0.06 ppm SO_2). The initial painting is assumed to cost \$0.50/square foot and subsequent paintings are assumed to cost \$0.40/square foot. To calculate the damage resulting from the more frequent painting of storage tanks, formula (5-1) is extended and applied(1). In a rural or unpolluted environment water storage tanks are painted every twelve years throughout their fifty-year life. The total present value of maintaining these tanks per square foot in an unpolluted environment is:

$$\$.50 + \frac{\$.40}{(1 + r)^{12}} + \frac{\$.40}{(1 + r)^{24}} + \frac{\$.40}{(1 + r)^{36}} + \frac{\$.40/6}{(1 + r)^{48}}$$

In a polluted environment the total number of repainting necessitated would be 50/8 or 5¼, and the total present value of maintenance would be:

$$\$.50 + \frac{\$.40}{(1 + r)^{8}} + \frac{\$.40}{(1 + r)^{16}} + \frac{\$.40}{(1 + r)^{24}} + \frac{\$.40}{(1 + r)^{32}} + \frac{\$.40}{(1 + r)^{40}} + \frac{\$.40/4}{(1 + r)^{48}}$$

1) The Fink *et al* study did not discount the costs of future repaintings and hence overestimates this damage. The calculation procedure presented here is the correct one.

The difference between these two yields an estimate of the present value of the damage due to extra painting because of air pollution. The value of this difference depends upon the chosen value for r which should be equal to the current interest rate used in social welfare projects. If, for example, a value of r equal to 0.05 were chosen, the present value of this difference would be $.30/square foot over the fifty-year period. As r increases this difference would decrease.

3. Estimating the monetary damage to materials

The component approach, as illustrated above, appears to be the best approach to estimate materials damage. It is much more difficult to analyse materials damage through the consideration of materials themselves. There are several uses of a given material and its value, lifetime and exposure to pollution are highly dependent on its use. This problem may be overcome through the use of mean value and average exposure levels, but there is some danger of losing sight of reality in carrying out such a task.

The first step in the component approach is a screening procedure using the I.S.I.C., or some alternative classification system, to determine those components susceptible to damage as a result of environmental degradation. In addition to those component systems used by Fink *et al* to estimate corrosion damage to metals, other (non-metal) components systems need to be defined. The M.R.I. study(1) provides a valuable list of those materials susceptible to air pollution damage.(2) The important non-metallic materials appear to be: synthetic rubber, natural rubber, plastics and related polymers, fibres (natural and synthetic) and paints. Major component systems incorporating these materials need to be defined In the case of specific pollutants the M.R.I. report indicates those materials which can be affected by given pollutants.

The component systems are defined and selcted on the basis of their economic importance, susceptibility to corrosion or effects from environmental changes and exposure to environmental changes. Those component systems which are most likely to be affected by other factors such as salt water and salt, before pollution effects

1) Salmon R.L. (1970), Systems Analysis of the Effects of Air Pollution on Materials, Final Report for Economic Effects, Research Division, National Air Pollution Control Administration, Raleigh, North Carolina, published by Midwest Research Institute, Kansas City, Missouri.

2) Reference is specifically made to air pollution because this is the type of environmental degradation most likely to yield financial losses to materials. In some rare cases water pollution could also yield material effects but there is little discussion of these effects in the literature.

can materialise, are likewise dropped from consideration in the screening procedure.

The second step is the collection of some data and information about the resulting component systems. Some idea of their number and value is needed. Estimation methods in this area are varied and highly dependent upon the component systems and upon data availability. Past shipment data from censuses of manufacturers is one source of data. The number of existing installations can be derived from estimates of new installations, making use of several assumptions. Given these results, an estimate of the amount of systems components exposed to air pollution (environmental degradation) is needed. This procedure again is highly specific to the component system studies. The study by Fink _et al_ for example, estimates the number of existing water storage tanks from supplier's data and from a knowledge of the total volume of water stored. The exposed surface area of the typical storage tank was then calculated and multiplied by the estimated number of tanks to arrive at the total amount of exposed surface. Of this total area some eighty per cent was assumed to be exposed to significant pollution.

The third step is the determination of the physical damage function, i.e. the specification of the physical effects of damage to the component system for the degrees of pollution (environmental degradation) considered. This is usually done through an examination of results from controlled experiments and on the basis of expert opinion. Fink _et al_, for example, look at the difference in replacement and repainting rates in rural and polluted areas to arrive at some estimate of the physical effects of air pollution. They consider only two levels of pollution in their physical damage function: the low level of pollution which generally characterises rural areas and some generally higher level which they estimate to be in the range of an annual average of 0.03 - 0.06 ppm of SO_2 in the case of water tanks, for example. The effect of this additional pollution is estimated to reduce the lifetime of the paint of the water tanks from 12 to 8 years. Other effects were considered unimportant in the case of water tanks. Similar specification of effects needs to be made for other material component systems. It should be stated that there is no need to restrict ocnsideration to two levels of pollution. The Programmes Analysis Unit(1) in the United Kingdom, considers three types of environments in its study of metal corrosion: rural, industrial and severe industrial. There is no need to limit oneself to three if there are sufficient data to consider additional degrees of pollution.

---

1) Programmes Analysis Unit (1971), _An Economic and Technical Appraisal of Air Pollution in the United Kingdom_, Chilton, Didcot, Berks.

The fourth step is the translation of the physical damage into monetary damage. In the examples given above, the use of formula (5-1) to calculate the additional costs due to premature replacement or repainting is demonstrated. There may be other financial losses associated with environmental deterioration. For example, the Programmes Analysis Unit finds that chemically resistant paint is often applied to steel work in severe industrial (highly polluted) environments. In this case, there are extra costs associated with a higher quality paint as well as with a greater frequency of painting. Both sets of costs need to be included in financial costs. Similarly, a study(1) of air pollution damage to rubber estimated the increased costs of producing air pollution-resistant rubber products by comparing the retail costs of rubber products, resistant and non-resistant to air pollution.

A final component of financial loss is that due to the necessity of using imperfect substitutes to avoid pollution damage. There appear to be no examples of this type of damage in the literature but it is conceivable that there could be some loss attributed to the decreased productivity, utility or desirability of a substitute product. Given the specific context, the decrease in productivity or utility can probably be calculated quite easily, although associated losses in amenity are difficult to estimate.

The final step in estimating the monetary damage to materials is relating the financial damage to the total monetary damage; i.e. estimating the loss of amenity component of the monetary damage. The major amenity loss considered in existing studies is associated with the costs of soiling. Several studies(2) indicate that despite the increased soiling associated with air pollution there appears to be little difference in the frequency of cleaning windows, buildings, etc. The financial costs of cleaning as a result of increased pollution are thus very small, but there may be social costs or losses in amenity resulting from dirtier windows and buildings. These costs are difficult to estimate but are bounded by the costs of cleaning. It can be argued that if the amenity losses were greater than the cleaning costs, then people would optimise their welfare by cleaning the soiled surfaces.

Except for damage to monuments, the other losses in amenity associated with the effects of environmental deterioration on materials are likely to be small. In the case of prized monuments the effects can be large and it is difficult to put a value on them.

---

1) Stickeney, Mueller and Spence (1971), "Pollution vs Rubber", Rubber Age, 45.

2) For example, Programmes Analysis Unit, op. cit. and Ridker, R.G. (1967) Economic Costs of Air Pollution, Praeger, New York.

Table 2

CALCULATION OF AIR POLLUTION DAMAGE TO WATER STORAGE TANKS IN THE UNITED STATES, 1970

Step 1:

Definition of component-systems susceptible to damage from environmental degradation.

Screening procedure beginning with I.S.I.C. or similar classification scheme.

Eliminate those component-systems which are:

a) rare or not important economically;

b) not susceptible to effects from environmental degradation;

c) not exposed to environmental degradation;

d) likely to suffer effects from other factors before environmental degradation can have any effect.

Result: Set of component-systems, example: water storage tanks

Step 2:

Data and information collection.

a) number and value of existing water storage tanks:

1\. i) A U.S. Dept. of H.E.W. report(1) gives an estimate of the total water stored in steel tanks as $11,000 \times 10^6$ gallons;

ii) Typical elevated storage tank contains 1×10^6 gallons (1MG);

iii) Estimated number of tanks 11,000MG/1MG = 11,000.

2\. i) Total tonnage of steel plate in all water systems is 7,100,000. 70 per cent of this is estimated to be in storage tanks;

ii) A 1MG storage tank uses about 350 tons of steel plate;

iii) Estimated number of water storage tanks (.70 x 7,000,000) ÷ 350 = 14,300.

3\. Final estimate; average of the above two estimates = (14,300 + 11,000) ÷ 2 = 12,650 water storage tanks.

b) quantity exposed to air pollution:

1\. Surface area of tank (exposed to atmosphere);

i) Hemisphere and roof area = 10,050 + 5,020 = 15,070 square feet;

ii) Other designs of water storage tanks have surface areas of from 12,000 to 30,000 square feet;

iii) Prototype tank was assumed to have a surface area of 20,000 square feet.

2\. Number of tanks exposed to air pollution: Water storage tanks tend to be located upwind of metropolitan areas, hence pollution exposure levels are not in general severe. 80 per cent of the water storage tanks are assumed to be exposed to average air pollution levels of from 0.03 - 0.06 ppm SO_2. The other 20 per cent are assumed to be exposed to little air pollution.

---

1) "Municipal Water Facilities - Communities of 25,000 Population and Over, United States and Possessions", A Report, U.S. Dept. of Health, Education and Welfare, Division of Water Supply and Pollution Control, Washington D.C., 20201, January 1964.

Step 3:

Determination of Physical Damage Function.

The only important effect is frequency of painting. Corrosion is generally avoided through protection by paint.

i) In rural (non-polluted) areas paint will protect steel for twelve years

ii) In urban areas with average pollution levels of 0.03 - 0.06 ppm SO_2, paint coatings typically last eight years.

Step 4:

Translation of the Physical Damage Function into the Monetary Damage Function.

1. The cost of preparing and painting a surface was estimated from several technical studies(1) to be \$0.50/square foot for the original painting and to be \$0.40/square foot for subsequent paintings.

2. The lifetime of water storage tanks was taken to be 50 years from U.S. Internal Revenue Service depreciation allowance data.

3. Calculation of monetary damage:

i) with no pollution (rural conditions), cost of repainting one square foot:

$$\$0.50 + \frac{\$0.40}{(1+r)^{12}} + \frac{\$0.40}{(1+r)^{24}} + \frac{\$0.40}{(1+r)^{36}} + \frac{\$0.40/6}{(1+r)^{48}}$$

where r is some appropriately chosen interest rate;

ii) with pollution, cost of repainting one square foot:

$$\$.50 + \frac{\$.40}{(1+r)^{8}} + \frac{\$.40}{(1+r)^{16}} + \frac{\$.40}{(1+r)^{24}} + \frac{\$.40}{(1+r)^{32}} + \frac{\$.40}{(1+r)^{40}} + \frac{\$.40/4}{(1+r)^{48}}$$

iii) the monetary damage from air pollution per square foot of exposed water storage tank surface is then the result in ii) minus the result in i);

iv) the present value of the total United States damage to water storage tanks from air pollution is the result in iii) multiplied by .80 x 12,650 x 20,000, the total surface area exposed to air pollution.

1) For example see Charlton, A.R. (1965), "The economics of Corrosion Control by Protective Coatings", Corrosion Technology, 12, pp. 11-16; and Brouillette, C.V. (1967), "Cost Comparison of Protective Coatings for Steel", NCEL Supplement, Technical Report R-50/-y-F020-03-01-001, NFEC.

Measures of willingness-to-pay (see below) may indicate the value of a Venice or Parthenon, but in any case, it appears best to represent the damage (or danger thereof) to these prized monuments in physical terms as a supplement to any monetary measure of the damage.

4. Special problems

There are two final points which should be raised in the estimation of materials damage. The first is that it is very difficult to associate material damage with a fixed, short time period of pollution, as a year for example. A single year of air pollution may have no effect on a painted surface; or, for accounting reasons, it may be exceedingly difficult to calculate the effect of one year's pollution at a certain level and a second year's pollution at another level, etc. For this reason, some ready-made assumptions are usually made in which the long-term effects of a certain fixed level of pollution over that period can be calculated. The result gives the effects over the entire time period and to obtain an annual result the result can be divided by the length of the time period. Hence, in the Fink et al study cited above, the damage costs of pollution to water storage tanks are calculated for a fifty year period. The present annual value of this damage cost was found by dividing the present value of the 50 years' costs by fifty.

The second point is that there is often no unanimous agreement on the input information and data used to calculate the financial loss. Where this is the case several sets of calculation can be undertaken for several sets of input data, thus leading to a set of final estimates of the financial loss. The advantage of this procedure is that if a single figure is presented as an estimate it is likely to be accepted as definitive. A set of figures tied to different sets of assumptions is less likely to create the impression that there is a definitive answer. The variability of final estimates is likely to be considered in any future decisions.

An alternative way of treating variable input information is that undertaken by the Programmes Analysis Unit. The authors find no agreement on the prices or value attached to materials affected by pollution, nor on the quantity and extent of the material damaged. To compensate for these divergences of opinion, they assume that existence of certain probability distributions for the input parameters and then make an estimate of the resulting financial loss along with a measure of the variability of this estimated loss.

As an illustration of this procedure let us consider the damage to agriculture from air pollution in the United Kingdom in 1971. This damage is estimated through use of the formula:(1)

1) This formula is derived in Appendix D.

$$C = E \frac{(F - 1)}{(F + R)} \qquad (5 - 2)$$

where E is the value of the national output of agricultural crops affected by pollution, F is the factor by which agricultural productivity in "clean" areas exceeds that in "polluted" areas, R is the ratio of the quantity of crops in "clean" areas to that quantity in "polluted" areas, and C is the agricultural damage from air pollution in 1971.

Agricultural experts were unable to agree on an estimate for E, the value of those crops susceptible to air pollution. The extreme estimates of this parameter were £200 millions and £2,300 millions per annum, yielding a median estimate of £1,250 millions. Results from field experiments and from comparing output in "clean" and "polluted" areas gave estimates for F between 1.01 and 1.20 with a median estimate of 1.1. R in this case was estimated to be 3.2. Using the various median estimates for E, F and R, a point estimate for C of £39 millions was made. To compensate for the variation in the estimates of the input parameters, probability distributions were assumed for these parameters (on the basis of the observed estimates given them). This made it possible to calculate the variability of the final estimate.

Where reasonable probability distributions are available for the input parameters, such an approach can be fruitful. In the absence of sound information about these distributions the measure of variability, however, should be utilised with caution as it is highly sensitive to the distributional assumptions.

D. MEDICAL EXPENDITURES

In the example in chapter III. A. the monetary damage associated with health effects consisted of three components: losses in wage income, increases in medical expenditure and psychic costs. It is thus clear that medical expenditures alone can never provide a complete evaluation of health damage, but they are an important component of the damage from health effects due to environmental deterioration.

From an individual point of view, the medical expenditure for a particular illness is well-defined. From society's point of view it is not so clear because individuals are generally protected by insurance. Moreover, medical care is generally not priced according to the marginal cost but in accordance with various social conventions. The total social cost for medical treatment due to environmental degradation is the opportunity cost of allocating resources to health care as a result of the degradation. Because there are many individuals and because there are insurance systems, the financial risks are generally pooled and it is not necessary to take any risk premium into account.

Estimates of medical expenditure are available for several categories of disease. These estimates have been applied to the estimated number of persons suffering as a result of reduced environmental quality. This procedure assumes that those suffering from a certain "disease" as a result of environmental degradation bear the same medical expenses as those suffering from the same diseases through other causes. There is not sufficient information available to evaluate this assumption.

In his estimate of the medical costs due to air pollution, Ridker(1) estimates the cost of treatment for chronic bronchitis in 1958. From government health statistics he is able to estimate the total number of patients who are medically attended, with and without bed-disability days. He is then able to make some assumptions for the average treatment costs for chronic bronchitis patients with and without bed-disability days. Multiplying these costs by the number of patients in each category, Ridker estimates the total national cost of treating chronic bronchitis in the United States in 1958. He then examines age adjusted death rates of chronic bronchitis patients for urban and rural areas and finds that there would be a total of twenty per cent fewer deaths if the rural death rate also prevailed in urban (polluted) areas. Hence Ridker multiplies the total cost of medical treatment for chronic bronchitis by twenty per cent to arrive at an estimate of the treatment costs for chronic bronchitis due to air pollution in the United States.

The interpretation of this estimate needs to be made clear. There are two levels of pollution under consideration, and the estimated cost is that which is due to the difference in these two general levels of pollution. The two levels are the past cumulative levels experienced in rural and those found in urban areas. The estimate is therefore based on several years of pollution experience and not simply one year. This fact must be taken into consideration in any use or interpretation of the result. A sudden decrease in urban pollution levels to those existing in rural areas would not erase the medical expenses completely. Those suffering from past pollution would continue to face medical costs. It is only in the long-run that the reduction would lead to zero. On the other hand, an immediate reduction in urban pollution levels would contribute to lower medical expenses extended over a period of several years. With better data than are currently available the present value of benefits from an immediate reduction in air pollution could be calculated, taking into account the two counteracting tendencies.

1) Ridker, R.G., *op. cit.*

Given the lack of data, it is reasonable to ignore these tendencies in the hope that they would cancel each other out. Table 3 gives a summary example of how medical expenditure cost estimates can be calculated from available information.

Allied to medical costs are the costs of premature burial. These are included in at least one study(1) of the costs of air pollution damage. These costs are the difference between the present costs of burial and the present value of the expected future costs of burial(2). The costs are rather small and can probably be ignored. This same reasoning should be applied to medical costs. To be correct, only additional medical expenses caused by reduced environmental quality should be considered. Hence the appropriate formula for calculating these costs is similar to the one used to calculate premature burial costs:

$$C_a = C_o - \sum_{i=1}^{m} \sum_{n=a}^{\infty} \frac{C_{ia}}{(1 + r)^{n-a}} \qquad (5 - 3)$$

where C_a is the present value of the increase in medical expenses due to environmental deterioration, C_o is the present cost of medical treatment caused by environmental pollution, C_{ia} = i^{th} deferred medical cost expected to occur n years from the present (a). These costs would have occurred except for the environmentally induced illness. For example, if the environmentally-induced disease leads to premature death, then the additional medical expenses must take into account those medical expenses otherwise expected.

It is conceivable that C_a could be negative, although the other damage components associated with illness are generally much larger than this component. In estimating medical costs, existing studies(3) have ignored the second term in (5 - 1). Relative to increased medical expenses, the result of this omission could be serious for those individual dying prematurely from diseases as a result of environmental degradation, where C_o is small; but relative to the total damage of health effects, this omission is not important.

E. WAGES AND LOST PRODUCTIVITY

The second health effects component of damage derived in the appendix to chapter III is that of lost wages. The wage loss is a

1) Ibid.

2) Jansen, H.M.A. et al (1974) Een Raming van de Schade Door Luchtvereininging in Nederland in 1970, Institut voor Mulieuvraagstukken, Vrije Universiteit, Amsterdam.

3) Jansen et al, op. cit., estimated this term. They found it to be small and feel that it can probably be ignored.

Table 3

CALCULATION OF MEDICAL EXPENDITURES FOR POLLUTION-CAUSED ILLNESS: BRONCHITIS, UNITED STATES, 1958

Step 1:

Identification of Effects.

Recourse to expert opinion and the results of epidemiological studies provided a list of deseases for which air pollution is, in part, responsible. Chronic bronchitis is among these diseases.

Step 2:

Relation of Physical Effect to Air Pollution Levels

1. Estimated number of chronic bronchitis patients. A United States government publication(1) reports that there were about 1,264,000 medically-attended chronic bronchitis patients between July 1957 and June 1958. Of this number, 595,000 had at least one bed-disability day in that period. Those patients of age ten or less were dropped from consideration because their condition is assumed to be independent of air pollution. The result leaves 416,500 bronchitis patients with at least one bed-disability day and 468,000 patients without bed-disability days.
2. Estimated portion of bronchitis cases due to air pollution. Results of a study(2) of adjusted bronchitis death rates in the United States indicate that bronchitis deaths would be reduced by 50 per cent if urban air pollution concentrations were reduced to rural levels.
3. Estimated total number of bronchitis cases due to air pollution. Assumption: the proportion of bronchitis morbidity due to air pollution is the same as the proportion of bronchitis mortality due to air pollution. Using this assumption, the estimated number of bronchitis patients with bed-disability days suffering because of urban air pollution is: 0.5 x 416,500 = 208,250 and the estimated number of bronchitis patients without bed-disability days suffering because of air pollution is: 0.5 x 468,000 = 234,000.
4. Translation of Physical Damage into Monetary Damage. The cost of treating patients with bed-disability days was assumed to be the same as the cost of hospital inpatients suffering from chronic bronchitis. The average cost of this treatment was given(3) as $169/bronchitis inpatient of which $41 was for attending physicians services. This amount was assumed to be the average cost of treating the bronchitis patient with no bed-disability days. The estimated total medical treatment costs for pollution-caused bronchitis in the United States in 1958 is then estimated to be: $169 x 208,250 + $41 x 234,000 = $44,788,250.

---

1) Department of Health, Education and Welfare, Public Health Service(1959), Health Statistics Chronic Respiratory Conditions Reported in Interviews, July 1957-June 1958, U.S. Government Printing Office, Washington D.C.

2) Lave, L.B. and E.P. Seskin (1970), "Air Pollution and Human Health" Science, 169.723-733.

3) Office of Dependent's Medical Care (1959). Dependent's Medical Care Program, Second Annual Report, Denver, Colorado.

direct private cost and only under certain conditions does it represent the social cost. The first condition is that the wage rate be equal to the marginal productivity of labour. Assuming a profit-maximising behaviour on the part of the employers and assuming that a single employer does not have the power to influence the wage rate, the wage rate will correspond to the marginal productivity even if the labour market is not competitive. The wage rate will then also measure the loss of production, provided that only a small number (small compared with the total labour force) of workers are absent because of pollution-induced illness.

If taxes are introduced into the discussion, the wage rate no longer represents private or social costs. The private cost is obviously the net earnings foregone minus any transfers which take place while the individual is on sick leave. The total social cost is equal to the production foregone and this is equal to the wage rate, plus any taxes paid by the employer.

Two types of lost productivity from pollution-induced ill health were distinguished in the appendix to chapter III: that due to absenteeism and premature retirement and that due to premature death. The first is measured through use of the income normally received in the period of absence. For non-remunerative services and activities, such as those of students and housewives, some value received for comparable work is usually chosen. The number of pollution-induced days of absence is usually determined by attributing some fraction (as determined by experts) of days absent, because of some specific disease, to environmental quality. The availability of the data necessary for this calculation varies greatly according to the study undertaken; they have been obtained from national health records, health surveys, industry records, and insurance companies.

The loss from premature retirement is usually derived through use of a formula similar to that of Ridker:(1)

$$V_a = \sum_{n=a}^{\infty} \frac{P_{a_1}^{\ n} \odot P_{a_2}^{\ n} \odot P_{a_3}^{\ n} \odot Y_n}{(1 + r)^{n-a}} \qquad (5 - 5)$$

where V_a is the present value of future earnings of an individual at age a, $P_{a_1}^{\ n}$ is the probability that an individual of age a will live to age n, $P_{a_2}^{\ n}$ is the probability that an individual of age a living at age n will be in the labour force at age n (if pollution-induced illness has not forced his premature retirement), $P_{a_3}^{\ n}$ is the probability that an individual of age a living and in the labour force at age n will be employed at age n, Y_n is the gross earnings at age n (unadjusted for consumption and taxes) and r is the rate of interest.

1) Ibid.

All of the above quantities can usually be estimated from demographic information or labour department statistics except for the interest rate, r, which will be dependent upon the assumptions which the authors of the study make. In existing studies r has varied from five to ten per cent.

This procedure implies that the average loss to a person suffering a pollution-induced illness is the same as that borne by the population at large with the same illness, not necessarily pollution-induced. This assumption may not necessarily be valid, as pollution may induce some illnesses at a younger age than generally experienced in non-pollution induced cases of the illness.

In the case of premature death, the loss to the society was shown to be the taxes the individual would have paid if he had survived and his net wealth accumulation if he had survived. This can be calculated in two ways. A value C_n, indicative of an individual's consumption at age n can be subtracted from Y_n to yield $Y'_n = Y_n - C_n$ and Y'_n can be placed in the Ridker formula (5 - 5) to calculate V'_a, the loss from premature death.

Alternatively, no allowance has been made so far for the loss of amenity associated with an increase in the risk of death. It can be argued that C_n is a lower limit for this loss and it can therefore be reinstated in (5 - 5). For premature death, V_a would then estimate the loss in productivity and some of the amenity loss of premature death. Table 4 presents an example of the calculation of this type of loss.

Table 4

CALCULATION OF LOST WAGES AND LOST PRODUCTIVITY

A. Absenteeism, bronchitis, United Kingdom 1970

Step 1:

Identification of Effects

See Table 3

Step 2:

Lost Productivity due to Air Pollution-Induced Bronchitis.

1.a) Department of Health and Social Security data indicate 32 million man-days of absence because of bronchitis. Because this estimate is based upon sickness benefit statistics, absences of less than four days are often not included, so this estimate was adjusted upwards to 41 million man-days lost because of bronchitis.

b) Department of Health and Social Security data show 4.5 million women-days lost because of bronchitis. Since three-fourths of married women are not insured under that National Health system this figure is adjusted upwards to 16.5 million lost women-days because of bronchitis.

c) Housewives are a large group not covered by the Social Security system. It is assumed that 14 million housewife-days are lost per year because of bronchitis.

2. Various studies indicate that the frequency of bronchitis cases is up to fifty per cent greater in urban areas than in rural areas. Because of the variation in these estimates it is assumed that the frequency of bronchitis cases in polluted areas is twenty-five per cent higher. As about 60 per cent of the population lives in relatively polluted areas, one can assume that about 20 per cent of all bronchitis cases (and absences) are due to air pollution.

Hence there are an estimated:
20 per cent x 32 million = 6.4 million man-days,
20 per cent x 6.5 million = 3.3 million women-days and
20 per cent x 14 million = 2.8 million housewife-days
lost because of air pollution-induced bronchitis.

Step 3:

Translation of Physical Damage to Monetary Damage

Taking the average daily wage paid to men (£4), that paid to women (£3), and extrapolating the female daily working wage to housewives, the total estimate of lost productivity from bronchitis induced by air pollution in the United Kingdom is
£4 x 6.4 million + £3 x 3.3 million + £3 x 2.8 million = £43.7 million.

B. Premature Death Chronic Bronchitis, United States, 1958.

Step 1:

Identification of Effects

See Table 3

Step 2:

Premature Death due to Chronic Bronchitis

See Table 3 where 50 per cent of all bronchitis deaths were assumed to be due to urban air pollution.

Step 3:

Calculation of Monetary Damage

1. The formula of Ridker(1) is used:

$$V_a = \sum_{n=a}^{\infty} \frac{P_{a_1}^{\ n} \ P_{a_2}^{\ n} \ P_{a_3}^{\ n} \ Y_n}{(1 + r)^{n-a}}$$

where V_a is the present value of the future earnings of an individual of age a and the other variables are as defined in the text.

$P_{a_1}^{\ n}$ $P_{a_2}^{\ n}$ and $P_{a_3}^{\ n}$ are calculated from age-specific data on life expectancy, labour force participation rates, and unemployment rates for the 1960 population. Y_n is the average wage of men or women aged n, depending upon whether the formula is applied to men or women. No adjustment is made for consumption, the result thus contains some part of the loss in amenity. Adjustment for consumption could be made by subtracting it from Y_n

2. The following sub-table from Ridker(2) applies this formula to chronic bronchitis deaths in the United States, 1958 using two alternative interest rates of 5 and 10 per cent.

(1) Age	(2) Number of Deaths	(3) Present Value of Earnings, Using Discount Rates of: .05	.10	(4) Total Loss .05	.10
a) Males					
10-14	7	$47,264	$17,826	$330,048	$124,782
15-19	4	59,800	28,035	239,200	112,140
20-24	3	69,676	37,690	209,028	113,070
25-29	6	73,143	43,098	438,858	258,588
30-34	7	70,523	43,982	493,661	307,873
35-39	11	64,038	41,834	704,418	460,174
40-44	23	55,638	38,129	1,279,674	876,967
45-49	30	45,836	33,077	1,375,080	992,310
50-54	84	35,001	26,690	2,940,084	2,241,960
55-59	150	23,419	18,920	3,512,850	2,838,000
60-64	269	11,685	9,917	3,143,265	2,667,673
65 69	314	4,056	3,559	1,273,584	1,117,526
70-74	268	692	651	185,456	174,468
Total	1,176			16,126,006	12,285,532
Average				13,713	10,447
b) Females					
10-14	4	13,391	5,377	53,564	21,508
15-19	4	16,809	8,351	67,236	33,404
20-24	4	17,794	9,434	71,176	37,736
25-29	5	18,408	10,390	92,040	51,950
30-34	9	18,927	11,666	170,343	104,994
35-39	8	17,946	11,907	143,568	95,256
40-44	15	15,136	10,543	227,040	158,145
45-49	18	11,822	8,600	212,796	154,800
50-54	20	8,653	6,651	173,060	133,020
55-59	31	5,357	4,345	166,067	134,659
60-64	39	2,411	2,030	94,029	79,170
65-69	54	847	743	47,196	40,122
70-74	62	134	125	8,308	7,750
Total	273			1,526,423	1,052,550
Average				5,591	3,855

The estimated total losses from premature chronic bronchitis caused deaths are $12,285,532 for men and $1,052,550 for women assuming a 10 per cent interest rate.

The estimated total loss from premature death due to air pollution-caused chronic bronchitis is then 50 per cent of the sum of the above two figures, or .50 (12,285,532 + 1,052,550) = $6,619,041.50

1) Ridker, R.G. (1967) Economic Costs of Air Pollution, Praeger, New York.

2) Ibid.

Chapter VI

ESTIMATION OF LOSSES OF AMENITY

A. SUMMARY

The methods for estimating losses of amenity are generally quite different from those methods used to estimate financial losses, although the direct method introduced in the last chapter can sometimes provide an estimate or an upper or lower bound for amenity losses. For example, the cost of cleaning is probably an upper bound to the amenity loss associated with soiling. Similarly, bounds or overestimates or underestimates for other amenity losses can be found through the use of direct methods, and these estimates can be the best estimates available. As the next chapter explains, such imperfect estimates can be useful and are sufficient for certain uses of damage functions.

The intangible aspect of amenity losses makes it much more difficult to define these losses. For this reason the methods which have been used to evaluate them are generally more sophisticated than the methods used to estimate financial losses. One natural method of estimating losses of amenity is to ask the individuals concerned through use of a questionnaire about the necessary compensation. Another method is to try to find some relationship between the loss of a specific amenity and the demand for a private good. Knowledge of this relationship may allow an estimate of the amenity loss. Sometimes these losses are capitalised in lower land values and a comparison of land values, either over time or over different neighbourhoods, may yield an estimate of the loss of amenity. Finally, litigation results may provide an approximation of amenity losses.

B. QUESTIONNAIRES

1. Definition and examples

Monetary damage has been defined as the amount of money necessary to compensate individuals for environmental deterioration

The most straightforward way to estimate the amount is to ask individuals directly about the monetary transfer which they consider necessary to compensate them. This approach suffers from two obstacles which make it difficult to interpret the responses to such questions. These are the problems of perception and of incentive.

The questionnaire approach is based on the assumption that it is possible to identify the group of individuals which is affected by the environmental deterioration. This is possible in many cases, particularly when only local changes in the environment are considered. If, for example, it is proposed to construct a plant discharging waterborne wastes in a village and if the only individuals who use the surrounding waters are the inhabitants of the village, it is sufficient to limit those questionned to the members of the village community. If, on the other hand, the village is also a tourist resort, the problem of indentifying the proper group to be interviewed is much more difficult. In some cases it is clear that the proper group is the whole population and the investigation must take the form of a special kind of referendum. It is possible, however, to use random samples and we will come back to this possibility later.

The responses to questions about the necessary compensation for environmental deterioration may be meaningless if the persons asked cannot perceive the change in the environment in the same ways as the investigators. In chapter II, we said that one way of overcoming the perception problem was to consider informed or 'expert' opinion. This was the approach adopted in the preceding section. Our justification for this approach is that in the long run public perception will closely reflect the perception of 'experts' trained to examine environmental effects.

In using questionnaires one way to overcome the perception problem is to combine the questionnaire procedure with a public education campaign. The latter may help offset the lack of public experience and knowledge with respect to given environmental effects. If a study were undertaken to evaluate the nuisance caused by increased noise levels from a new airport, it may be meaningless to ask the public to evaluate the noise nuisance. The population, in general, has not experienced such noise nuisance and cannot easily appreciate it. The problem could be at least partially overcome if the effects were formulated very carefully by, for instance, equating noise levels to disturbance to conversation, television viewing, sleep, etc., i.e. if the public to whom the questions are posed is informed and taught about the effects of the potential noise.

Even where there is an education campaign there is evidence that individuals find it difficult to evaluate a hypothetical situation, and their evaluation remains highly sensitive to the way in which the question is posed. For example, in the study undertaken to help determine a site for the third London airport(1) people were surveyed in an effort to estimate the social costs of moving (because of noise). They were first asked the following question: "Suppose you discovered that house prices in this area were higher than prices in similar areas five or more miles away; say that in those areas you could get a house comparable to this one and make £100 profit after all expenses, would you seriously consider moving? <u>If NO</u> what would the difference in price have to be to make you seriously consider moving?" This question was used to obtain an estimate of the consumer surplus attached to a home. At first, thirty-eight per cent of the respondents indicated that no amount of money would induce them to move; i.e. thirty-eight per cent indicated that they had infinite consumer surpluses attached to their homes. The Roskill Commission research team conducted a similar survey, but with a differently worded question. They found only eight per cent 'infinite'. The difference in results between the two surveys indicates that public understanding or perception of an environmental situation is highly dependent upon the question posed. In the second case people were given the impression that they were compelled to sell their homes and this probably has led to a distorted response.

Even when the public has sufficient experience there is some doubt as to whether damage can be properly appreciated or given a monetary value by a public which is not accustomed to evaluating. For example,(2) immediately after an unusual and severe soiling incident, residents in Syracuse, New York, were asked how much they would have been "willing to pay to have kept this incident from affecting" them. At the same time supplementary information was obtained which allowed an estimate of the actual expenses incurred by the residents. Of the eighty-five individuals sampled only eleven appeared to answer the first question rationally(3) (i.e. they were willing to pay at least as much as their expenses to avoid the incident). The inconsistency in the responses reflects

---

1) Plowden, S. (1970) <u>The Cost of Noise</u>, Metra Consulting Group Ltd., London, and Commission on the Third London Airport Report, (1971) HMSO. For a further discussion of these techniques and others connected with the estimation of noise nuisance from a third London airport, see Alexandre, A. and J-Ph. Barde (1973) <u>Le Temps du Bruit</u>, Flammarion, Paris and Dasgupta, A.K. and D.W. Pearce (1972) <u>Cost-Benefit Analysis: Theory and Practice</u> Macmillan, London.

2) cf. Ridker, <u>op. cit</u>.

3) Perhaps the responses are indeed rational and people derived some vicarious pleasure from the soiling incident. It may have altered an otherwise monotonous life pattern and provided a subject of lively conversation.

an inability to perceive the damage correctly or to place a monetary value on the damage. It is unclear whether a reworded question or an altered interview technique would have led to appreciably better results.

2. Variations of methodology

Delphi techniques can be included among questionnaire methods. The method is essentially an iterative questionnaire procedure involving a group of 'experts' and its success will depend largely upon the questions posed and the knowledge of the 'experts'. There is, in addition, the problem that 'expert' perception may not be the same as 'public' perception.

There are some interesting experiments which suggest that sophisticated variations of the questionnaire approach may yield good estimates of amenity losses. Foremost among the pioneers with these methods have been Hoinville and Prescott-Clarke(1). They have presented subjects with a series of scaled alternatives for which the subjects were forced to express their preferences, trading-off one alternative against another.

In the particular study cited, the trade-off is essentially between shorter and longer travel time on the one hand and the volume of traffic passing a subject's home on the other hand. The reasoning behind this trade-off is the fact that any measures taken to reduce the volume of traffic in residential areas would probably lead to the implementation of new traffic patterns which would increase traffic congestion elsewhere and thereby lengthen driving time. Ideally, most persons(2) would prefer minimal travel time and traffic volumes. The objective is to try to determine some value to attach to these preferences.

In the experiments undertaken, some 850 subjects were polled as to their preferences. The first step was to ensure that all subjects clearly understood the alternatives available.

Two approaches are in use to evaluate the preferences. The first is a procedure in which the subjects indicated the size of a weekly tax which they would pay to achieve a given improvement in traffic volume or travel time. Amenity levels were carefully explained for each of the alternatives and subjects were told they would have to pay a weekly tax of 10p. per level of improvement. Hence to reduce the level of residential traffic from the '0-level' (continuous traffic: mainly lorries) to the '2-level' (moderate traffic: some lorries), subjects would need to pay a weekly tax of 20p. The results of this approach are presented in table 5.

1) See, for example, Hoinville G. and P. Prescott-Clarke (1972) Traffic Disturbance and Amenity Values, Social and Community Planning Research (SCPR), London.

2) Ninety per cent of those participating in the survey.

Table 5

AVERAGE AMOUNTS SPENT TO IMPROVE RESIDENTIAL TRAFFIC VOLUMES

I In weekly tax — Traffic Volume in Residential Neighbourhood

	Total Sample	Moderate Traffic: Mainly Cars	Moderate Traffic: Some Lorries	Continuous Traffic
Total group	6p	3½p	8p	15p
Those who would like reduced traffic volume	10p	5p	9p	16p
Those willing to pay for improvement	19p	10p	17p	28p

II In weekly travel time

	Total Sample	Moderate Traffic: Mainly Cars	Moderate Traffic: Some Lorries	Continuous Traffic
% trading-off travel time to improve amenity	33%	28%	43%	68%
Existing residential score	2.6	3.0	2.0	0.5
Average improvement made to residential amenity score	0.4	0.3	0.6	1.2
Average minutes/hours sacrificed	4 mins.	3 mins.	6 mins.	12 mins.
Average total road time/week	3.6 hrs.	3.9 hrs.	3.2 hrs.	3.2 hrs.
Average minutes sacrificed/week	14½	11½	19	38½
Value of trade-off/week (at 24p/hr of travel)	6p	4½p	7½p	15½p

For the second approach, called the 'priority evaluator approach', respondents would have to pay for improvements in the volume of residential traffic through sacrifices either in accessibility or travel time. All of the subjects in this experiment travelled frequently in their personal vehicles and were given moderate accessibility ratings of '2'. Depending upon the existing traffic volume in their residential neighbourhood, residents were given additional points. Thus a person who experienced moderate traffic - mainly cars, was given three points for residential traffic volume amenity and two points for the amenity of accessibility (travel time) for a total of five points. The subjects then re-allocated their total points to achieve a personal optimal balance between amenity and accessibility. Hence, the above subject with five points might choose to sacrifice some accessibility by increasing his travel time 10 minutes per hour in exchange for augmentation in amenity to the point where there would be very little traffic in his neighbourhood. In this case he would redistribute his five points, allotting four instead of three to amenity and one instead of two to accessibility. Table 5 shows the average amount of time sacrificed to achieve an improvement in amenity.

If one accepts the conversion rate of 24p per hour of travel time, the two approaches give comparable answers. The average hourly earnings of manual workers in the U.K. are much higher(1) than the 24p used in this study. It is unclear if the value to be placed on travel time should be equivalent to the wage rate. If this equivalence is correct (or even if the value of travel time were equal to one half the wage rate), then the approach of asking people about their willingness-to-pay may have significantly underestimated the value attached to the amenity of decreased residential traffic volume.

The priority evaluator approach for estimating the values attached to amenity benefits (or losses) may have substantial promise in the future. Subjects may be more capable of relating amenities to each other through the making of trade-offs than of attaching monetary values directly to the amenities. It remains only to choose as a gauge an amenity (such as travel time) to which some monetary value can be attached. This choice and the attachment of a monetary value to it may not be easy as is seen in the Hoinville-Prescott-Clarke experiment. There are additional problems associated with the application of this method. The nature of questions and responses remains hypothetical, and it is difficult to discern if the environmental states perceived by the subjects are the same as those they are intended to perceive. Finally there are considerable resources required in designing and implementing this approach.

3. Problems of deliberate distortion of responses

Another problem associated with the questionnaire approach is the incentive for individuals to distort their responses. Independent of whether income transfers actually take place between the individuals and the authority, the individuals can have strong incentives for concealing their true preferences. As an example,

---

1) From the Department of Employment Gazette, March 1973, Volume LXXXI, No.3, Her Majesty's Stationery Office. The average hourly wage for full-time employed men (21 years and over) for all major industries was 79.6p in October 1972. The corresponding average hourly wage for women (18 years and over) was 48.28p.

assume that it is proposed to improve the water quality in a given lake by two means: first, by the direct cleaning of the lake, the cost of which will be borne by the community in the form of tax increases; secondly, by the reduction of effluent discharges into the lake, the cost of which will be borne by the firms using the waste disposal capacity of the lake. In order to decide to what extent this improvement should be carried out, an estimate of the benefits of the improvement is desired. We assume that members of the local community will be the only beneficiaries of any improvement in the water quality of the lake.

The benefit from the improvement (or the reduction in the monetary damage) is, by our definition, the amount of money it is possible to extract from the members of the community without putting them in a worse situation. This amount is the maximum willingness-to-pay. The purpose of a questionnaire is to estimate this willingness-to-pay.

It is in the interest of those responding to overstate their willingness-to-pay if the responses to the questionnaire carry no financial obligations. As the amount they actually will pay is independent of their responses to the questionnaire, it is in their interest to promote the considered quality improvement in the environment as much as possible. This can be done by overstating their willingness-to-pay. The only relationships between their responses and the increased amount they would actually pay is the expected increase in taxes. Only a part of the total costs will fall upon the individual in terms of this tax increase as the costs of improvement will be diffused throughout the community. For this reason it is not possible to take the responses at their face values and they can grossly overestimate the true benefit of the improvement.

If, on the other hand, financial obligations are dependent on the responses, it is in the interest of the individuals to understate their willingness-to-pay. If an individual believes that he will have to pay an amount equal to his stated willingness-to-pay, he will probably state a willingness-to-pay smaller than the true one because he may hope that the other individuals will contribute sufficiently for the equality for the lake to be improved. This is an example of the 'free rider' problem.

In theory it is possible to design the financial obligations in such a way that the incentives to misrepresent the true preferences are minimised. If, for example, the individuals believed that they must pay a certain fraction of their stated willingness-to-pay, this problem could be overcome. If the fraction were close to one, the free rider problem would prevail and the true willingness-to-pay would be underestimated. If, on the other hand, the fraction were close to zero, the individuals would find quality improvements to be very inexpensive, and they would overstate their willingness-to-pay. Thus, it seems clear that there must exist some value between zero and one, where these two counter incentives balance each other. The difficulty is that the true willingness-to-pay needs to be known in order to compute the correction fraction.

In general it is possible to design questionnaires in various ways so that financial obligations assume many different forms. For most of these ways it is possible to learn whether the incentive to overstate the true willingness-to-pay is greater than the incentive to understate. This makes it possible to undertake a study using different designs for different samples of individuals from the same population, instead of uniformly polling the entire population. A design that is known to give understatements can be applied to one random sample at the same time that a different questionnaire design, known to give overstatements, is applied to a second random sample. The two sets of results yield an interval which contains the true willingness-to-pay with a certain probability. This information can be sufficient to make a rational decision possible.

Although this approach has not been applied to the estimation of environmental benefits, it has been applied to other problems, such as the evaluation of television programming(1). We will not discuss the empirical results here, but refer the interested reader to the references given below.

The principal difficulty with this approach, however, is that it implies that different financial obligations are imposed upon different segments of the community. This may not be feasible from an income distribution point of view. Furthermore, this technique presents financial obligations in the form of a tax, and in no country does reality correspond to a situation in which taxes can be individually adjusted as assumed in the above approach.

4. Overall evaluation

Despite the overall discouraging experience in the use of questionnaires to estimate amenity losses, some of the newer techniques as the priority evaluation approach, indicate a great deal of promise. If these methods can be tested and developed further they may constitute one of the most important approaches in the

1) See Bohm, P. (1971) "An Approach to the Problem of Estimating Demand for Public Goods", Swedish Journal of Economics, 55, Bohm, P. (1972) "Estimating Demands for Public Goods: An Experiment", European Economic Review, 3, pp. 111-130, and Kurz, Modercai (1973), "Experimental Approach to the Determination of the Demand for Public Goods", NSF Grant Series, Working Paper III, Institute for Mathematical Studies in the Social Sciences, Stanford University, Palo Alto, California.

future. Questionnaire methods have the advantage of greater direct public participation in the evaluation of priorities and damage; moreover, samples of those questioned can be stratified by income, which facilitates any adjustment for income distribution purposes.

C. CONSTRUCTION OF DEMAND CURVES FROM INFORMATION ON INDIVIDUAL BEHAVIOUR

1. Background and rationale

Observation of individual behaviour can sometimes allow the deduction of monetary damage through the construction of individual demand curves. This approach underlies the above discussion of financial losses, but here we will apply it specifically to the estimation of losses in amenity.

The fundamental assumption here is that there are relationships between physical damage or environmental quality and the demand for pure, private, marketable goods. Through an examination of the demand for some private good, it may be possible to draw some conclusions about the willingness-to-pay for environmental improvements. The relationships between environmental quality and the demand for certain goods can vary greatly, ranging from complete complementarity between environmental quality and the private good, to a relationship of competitive goods. The basic difficulty is the advance specification of the probable relationship. This is a task which often requires much imagination and talent. Once the relationship is specified, it is possible to compute the monetary damage from information on individual behaviour. We illustrate this approach for three types of relationship: complete complementarity, complete substitutes and weak complementarity(1).

Complete complementarity between two goods is defined by the requirement that the two goods must be used in fixed proportion to each other. The classical textbook example is the case of a left shoe and a right shoe. One shoe is worth nothing to the consumer without the other, nor is an extra shoe worth anything to a consumer. More meaningful examples of complete complementarity particularly as regards environmental quality, are difficult to find. The purpose of this discussion is primarily to introduce the concept of weak complementarity.

Let us assume that there is an environmental service, which is supplied in quantity Y. Corresponding to this service, let it be

1) For a systematic discussion of these relationships, see Mäler, K-G. (1974) Environmental Economics: A Theoretical Inquiry, The Johns Hopkins Press, Baltimore, and Sandmo, Agnar, "Public Goods and the Technology of Consumption", Review of Economic Studies.

assumed that there is a marketable good, demanded in the amount X. The assumption of perfect complementarity implies that there exists a constant a, such that if $Y > aX$, there is an excess supply of the environmental service, $Y - aX$, which is of no value to the consumer; likewise, if $Y < aX$, there is an excess of the marketable good. It is clear that the latter situation cannot prevail in equilibrium because the consumer would reduce his consumption of X to the level: $\frac{1}{a} Y$. The former situation can, however, prevail in equilibrium because Y is a public good and the consumer has no direct influence on the decision to supply public goods. If this situation does prevail, the consumer obviously would not be willing to pay anything for an increase in the supply of the service and it is not necessary to compensate him for a small reduction in the service.

The units of the various quantities can always be defined so that the constant a is equal to one. In the following discussion it is assumed that a = 1.

As X is a private marketable good, there exist demand curves for X, which are in principle observable. In order to simplify the following discussion, we neglect the distinction between Marshallian demand curves and income-compensated demand curves, although it is possible to incorporate this important distinction into the analysis, at the cost of a more complicated presentation. The demand curve for X is drawn in Diagram 4.

Diagram 4

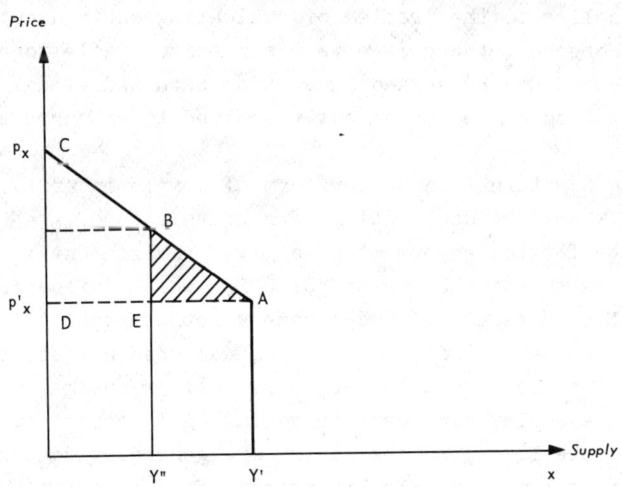

Let us assume the price of the marketable good is p'_x and that the supply of the environmental service is Y'. The consumer will then be in equilibrium at point A in the diagram. At this point there will be neither an excess of the private good nor cf the environmental service. If the supply of the environmental service is reduced to Y", the new equilibrium will be at point B. When the supply of the service is at Y' the consumer surplus is equal to the area of ABCDEA; in the new equilibrium, with supply at Y", the consumer surplus is equal to the area BCDEB. The loss in consumer surplus is therefore equal to the area of the shaded triangle ABE. The monetary damage to the individual from this reduction of the environmental quality is equal to this triangle. An increase in the supply of the environmental service, that is, an environmental quality improvement, would not change the consumer surplus, because the price of the private commodity will restrict its consumption to Y'. When the initial situation is at A, a quality improvement would thus yield no benefits.

Let us assume that the initial situation is characterised by the price p'_x and by the environmental service Y". The equilibrium is now at point B, and the environmental quality is the scarce factor. In this situation, a quality improvement would yield benefits. If the supply of the environmental service would increase from Y" to Y', the associated benefit would equal the shaded triangle ABE.

If we know the demand curves (or the income compensated demand curves), we can compute the monetary value of environmental changes when there is a perfect complementarity between environmental services and private marketable goods. Once again it must be stressed that the assumption of perfect complementarity very rarely, if ever, applies to the problem of evaluating environmental quality changes. Moreover, there is a very serious obstacle to estimating the income-compensated demand curves for each individual because the above method cannot be directly applied to an aggregate demand curve(1).

Let us now turn to a weaker form of complementarity: those situations where the consumption of a private good is a necessary prerequisite for the enjoyment of a given environmental quality. An example is the case of water pollution which influences the use of a lake for recreation. Under some circumstances it appears reasonable to assume that if an individual does not use the lake for recreation, he is not concerned by quality change in the lake. As a second example, consider air pollution in a particular region. Only the people living in the region are generally concerned with its air quality; most people living outside the region are indifferent. There are private marketable goods associated with environmental equality in both examples; in the first example recreation

1) The price at which there is no excess supply of either good will most likely differ among individuals. The aggregate demand curve will therefore be unlike that in Diagram 4.

is the marketable good, in the second, it can be certain properties. If an individual does not consume this private good in general, he is usually indifferent to the environmental quality changes concerned. This assumption does not hold, however, when option values are involved or when individuals demonstrate positive willingness-to-pay for quality improvements whether or not the improvements concern their own environment.

As an example, we might consider recreation at a certain lake as a private good. It is, in principle, possible to estimate the demand function for this good. The amount of recreation demanded at this lake will depend upon the price of recreation (including transportation costs, possible fees, equipment, etc.) and the water quality of the lake. Let us assume that the initial water quality is equal to Y'. We can now illustrate the demand function by the demand curve AB in Diagram 5.

Diagram 5

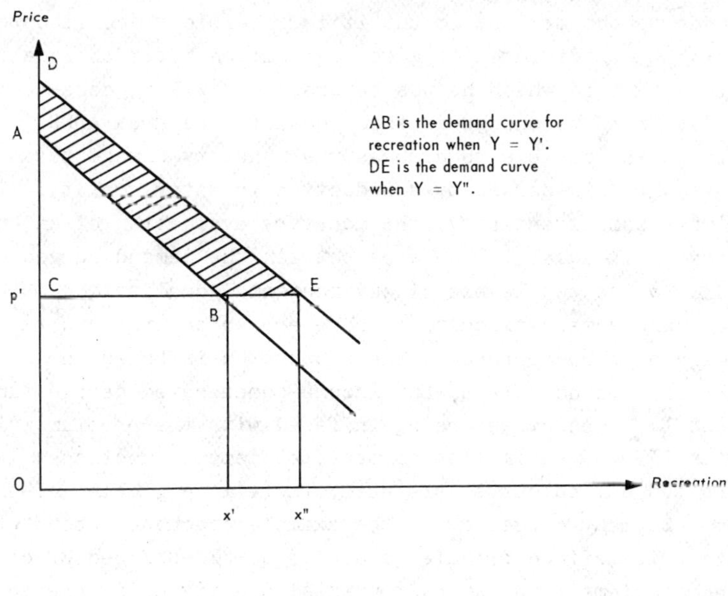

If the water quality were to improve to Y" the demand curve would shift to DE. Let us assume that the price of recreation is equal to p' or OC in the diagram. The individual will then demand the quantity x' of recreation (measured, for instance, in recreation days). When the quality has improved, the individual will increase his demand to x". The problem is to determine the quantity the individual is willing to pay for this increase in quality. To solve this problem, let us divide the movement from the point B to the point E into three steps:

a) assume the price increases from OC to OD. This would imply that the individual will demand no recreation and his loss in consumer surplus is equal to the area ABC;

b) if the water quality improves from Y' to Y", the demand curve will shift to the position DE, but as there is no recreation consumption at point D, the individual, according to the assumption of weak complementarity, will be indifferent to the quality improvement;

c) if the price decreases from OD to its previous level OC, the consumer will demand x" units of recreation and his consumer's surplus will increase by the area ECD.

In order to compensate the individual for the price increase of the first step, it is necessary to give him his loss in consumer surplus, the area ABC. As he is indifferent to the quality change in step b), no compensation is needed. Finally, in the last step, it is possible to confiscate the amount ECD without putting the individual in a worse position at E as compared with D. The net transfer is thus equal to the area ADEB, and this net transfer corresponds to the maximum amount it is possible to confiscate without placing the individual in a worse position after the quality change than that in which he was before. His willingness-to-pay or his benefit from the change is thus equal to the area ADEB. Similarly it is possible to discuss the monetary damage to an individual which is caused by a reduction in water quality. In the case of weak complementarity, the monetary evaluation of an environmental change is equal to the area between the demand curves which correspond to the two levels of environmental quality.

There are some difficulties, however, which arise in the application of this approach. The discussion is based on (Marshallian) demand curves, but income-compensated demand functions are needed(1). Hence, we are again faced with the problem, discussed in chapter III, of estimating compensated demand functions. Even if uncompensated demand curves are used, there is a problem of estimating them. In many cases, as in the example discussed above, there exists no well-defined private good with a well-defined price which is a weak complement to the environmental quality of interest. Recreation means different things to different individuals, and there is no general price which can be attached to this activity. The solution to this problem must involve a careful analysis of the recreational activities that take place in the lake and a study of

1) For a complete discussion, see Mäler, K-G. (1971) "A Method of Estimating Social Benefits from Pollution Control", Swedish Journal of Economics, 73(1), and Mäler, K-G. (1974) Environmental Economics: A Theoretical Inquiry, Johns Hopkins Press, Baltimore. (Chapter IV).

the costs involved in undertaking these activities. Examples of this approach are given below. So far we suggested how the demand function can be constructed as a function of price. It is also necessary to determine the relationship between the demand function and environmental quality. This is a problem which is very difficult to discuss in general, but we will give some examples below in which this problem has been treated.

Before considering these examples, let us consider the situation where an environmental service is a perfect substitute for a private marketable good. In some cases, private swimming pools may be regarded as substitutes for clean lakes and streams. In other cases air conditioning may be regarded as a substitute for clean air. There appear to be several situations where an environmental service approximates to being a perfect substitute for a private good, and it is this observation which underlies most of the estimates of financial losses which were discussed previously in this chapter.

Let us assume that we have a pair of perfect substitutes, one of which is a private marketable good and the other an environmental service. An increase in the supply of the environmental service will induce the individual to reduce his consumption of the private good by a certain amount. As the two goods are perfect substitutes, this substitution does not change the well-being of the individual except that it is now possible for him to achieve the same satisfaction as before in a less expensive way. Thus, the reduction in expenditure on the private good is exactly the amount it is possible to confiscate without leaving him in a worse situation. (The distinction between compensated and uncompensated demand curves is ignored here also.)

In situations where a private good is a perfect substitute for an environmental service, alterations in expenditure on the private good as a result of environmental changes can be used as an estimate of the monetary damage or benefit. The principal difficulty with this approach is the identification of those situations.

An example where a private good approximates to being a perfect substitute for an environmental service is taken from a recent study(1) which estimated residential amenity losses from increased noise levels. The study examines the cost of noise disamenity through consideration of a model which simulates the decision process of home owners on whether to move or not to move.

In its simplest form the model indicates when a person will tend to move or stay depending on whether N is greater or less

1) Commission on the Third London Airport (1971), Report HMSO, London.

than S + D + R, where N is the disamenity cost of noise, S is the householders' surplus before any change in his environment, D is the house price depreciation due to increased noise and R is the moving cost.

Starkie and Johnson(1) further refine this model by breaking up N into two components G and N'. G is the cost of sound insulation in homes and N' is the residual amenity loss, N - G. In adopting this approach they are assuming that the insulation costs are, in part, a perfect substitute for a quiet environment. The quantity of money spent on insulation costs is dependent upon the peacefulness of the environment and can be used to estimate the monetary damage of the introduction of noise. In this particular case, the result will be an underestimate (by the quantity N') because insulation is not a perfect substitute for all aspects of environmental tranquillity. Windows open to let in fresh air, for example, and gardens, cannot be insulated. However, Starkie and Johnson are able to estimate G, and they believe they can estimate N' through an examination of empirically-derived figures of property turnover in noise-affected areas.

Amenity losses are often associated with the loss of recreation or of the possibility of assuming some private good because of decreased environmental quality. This situation corresponds to the weaker form of complementarity and the general approach towards estimating amenity loss is provided above. The first step is the estimation of the demand curves for the private good as a function of environmental quality.

2. The simplest case: two levels of environmental quality

Several studies have estimated the demand for a resource, particularly recreational resources. The simplest case is that in which there are only two possible states of environmental quality:

1. That in which the demand is zero, and;

2. An acceptable quality state where the demand is non-zero.

In this case, the change in environmental quality is that which reduces an existing demand curve to zero or vice versa. There are no intermediate demand curves. This situation prevails when environmental changes make an existing resource unavailable for future use or when a previously non-existent resource becomes available.

Here the demand curve AB in Diagram 5 passes through O, and it is necessary only to estimate the demand curve DE. The benefit

1) Starkie, D.N.M. and D.J. Johnson (1973) "Losses of Residential Amenity: An Extended Cost Model", Journal of Regional Studies, 7(2).

(or damage) then corresponds to the area under the demand curve DE. We shall begin with a discussion of this case and will later extend it to those cases where there is a continuum of environmental qualities. We shall also introduce the additional complication of competing resources below.

Several approaches have been used to estimate the demand curve DE. Foremost among them is that of Clawson and Knetsch(1). For a given natural resource, they construct a demand curve through use of surveys which estimate the total cost to parties visiting the resource and the number of parties visiting the resource at each level of cost. The principal component of this cost is travel expense.

The methodology is illustrated in Table 6. The first step consists of constructing a set of concentric circles around a recreational area. The areas between the circles define distance zones. For each zone the total population and the number of visits by that population to the resource are obtained, thus allowing a calculation of the number of visits per 1,000 base population. The resultant visit rate is plotted against the estimated cost per visit for a party of four to obtain an estimate of the demand curve. Since the larger part of the estimated costs are due to travel expenses, the grouping of visitors by the distance they live from the resource is reasonable. Clawson and Knetsch suggest that several groups be defined in order to obtain a demand curve whose shape is well-defined. A further refinement would be to define groups in terms of both distance lived from the resource and income and perhaps also according to the use made of the resources (swimming, fishing, hiking, etc.).

The Clawson-Knetsch estimate gives the total demand for the current year. The total benefit of a resource is the sum of the current year's benefit and the benefits in future years. The calculation of the latter involves assumptions about future population levels and visitation rates, as well as an assumption about the interest rate. The visitation rate is likely to be a function of the population, but the limited capacity of a resource must not be ignored. Increased visitation rate and population growth can be combined in a third parameter, α; the annual growth rate of the benefit. The present value of the yearly benefit a years from now (B_a) can then be estimated as:

$$B_a = \frac{B_o (1 + \alpha)^a}{(1 + r)^a} \qquad (6 - 1)$$

1) Clawson, M. and J.L. Knetsch (1971) Economics of Outdoor Recreation, Resources for the Future Inc., the Johns Hopkins Press, Baltimore.

where B_o is the present yearly benefit, α is the annual growth rate of benefits and r is the interest rate.

Because the capacity of a resource is limited, $(1+\alpha)^a$ is bounded by the capacity. Where $(1+\alpha)^a$ is greater than the capacity, A, the yearly benefit B_a is $\frac{B_o A}{(1+r)}a$. If the value of α differs among the zones, the above formula can be used to calculate the specific future benefits for each zone, and these benefits can then be summed over zones.

The establishment of a new resource will, to some extent, shift demand away from existing resources. The calculation of the benefits of the new resource may, in such a case, be more difficult. If, however, the price (or cost) of using the existing resources is unaffected by the new resource, and if the demand curve for the new resource is estimated on the basis of this assumption, the consumer surplus is still given as the area below the demand curve for the new resource(1).

---

1) In Appendix A the benefit from an improvement in environmental quality was written as:

(15) $B^h = I^h - m^h\ (P, Y'', A)$

If we apply this formula to the present case, we get:

$$B = m(\bar{\bar{p}}_1, \bar{p}_2, A) - m(\bar{p}_1, \bar{p}_2, A)$$

where $\bar{p}_1$ is the cost of using resource 1, $\bar{p}_2$ the cost of using resource 2 and $\bar{\bar{p}}_1$ a price on resource 1 sufficiently high so that the demand is zero.

The situation before the new resource is opened up is then characterised by $\bar{\bar{p}}_1$, $\bar{p}_2$ and this situation is denoted A. When the new resource is opened up, the cost of using that resource is $\bar{p}_1$.

B can now be written:

$$B = \int_{\bar{p}_1}^{\bar{\bar{p}}_1} \frac{\partial m(p_1, \bar{p}_2, A)}{\partial p_1}\, dp_1$$

It was shown in the appendix, however, that:

$$\frac{\partial m}{\partial p_1} = x_{,}(p_1, \bar{p}_2, A)$$

where $x_1(p_1, \bar{p}_2, A)$ is the compensated demand function for resource 1. If we identify the compensated demand curve with the Marshallian demand curve, we finally get:

$$B = \int_{\bar{p}_1}^{\bar{\bar{p}}_1} x_1(p_1, \bar{p}_2, I)\, dp_1$$

The benefit is given by the arc under the demand curve for the new resource.

Table 6

CALCULATION OF BENEFITS OF KERR RESERVOIR, NORTH CAROLINA AND VIRGINIA

A. Population, Party Visits and Yearly Visit Rates to Kerr Reservoir, North Carolina and Virginia, 1963-64, by Zone of Origin of Visitors.

Mileage limits of zone of origin	1960 population (1,000)	Total party(1) visits	Visits per 1,000 population %	Two-way travel cost at 5.16¢ per mile
1- 20	91	165,104	1,814.32	$ 1.34
21- 40	97	60,338	622.04	3.10
41- 60	225	56,284	250.16	5.16
61- 80	642	37,606	58.58	7.22
81-100	875	19,600	22.40	9.29
101-140	2,503	46,869	18.73	12.38
141-180	1,245	17,213	13.83	16.51
181-220	2,693	11,914	4.42	20.64
221-260	3,795	3,541	0.93	24.77
261-300	2,861	2,327	0.81	28.90
301-400	15,141	1,925	0.13	36.12
401-500	29,037	2,529	0.09	46.44

1) Number of vehicles

Source: Calculated from unpublished surveys of the U.S. Corps of Engineers

B. Calculation of Benefits(2) for 1963-64 Visits

$$B_{63-64} = \sum_{zone\ i=1}^{12} Visits_i \times Travel\ Costs_i$$

$$= \$(165{,}104 \times 1.34 + 60{,}338 \times 3.10 + \ldots + 2.529 \times 46.44)$$

C. Calculation of Future Benefits for Year 1963-64 + a.

$$B_{63-64+a} = \sum_{zone\ i=1}^{12} (\text{population in Zone } i)_{63-64+a}$$

$$\times (\text{Visit rate zone } i)_{63-64+a} \times \frac{\text{travel-cost zone } i}{(1+r)^a}$$

where r is the rate of interest. If, for example, the expected population in 1983-84 in zone 1 is 180,000 people and the expected visit rate (per 1,000 population) remains at 1,844.33, the first term in the above sum will be:

$$180 \times 1{,}814.33 \times \frac{1.34}{(1+r)^{10}}$$

D. Total Benefit

The total benefit is the sum of the benefits for the individual years calculated in part C. As $(1+r)^a$ becomes large, the contribution of year 1963-64 + a will become small and after some future year it can be ignored.

2) In 1963-64 dollars.

If the introduction of a new resource affects the congestion (quality levels) of existing resources, it is necessary to add the benefit from reduced congestion to the above benefit. This can be done through use of the methods described below in section 3.

The use of travel costs as a price for the use of a resource can be criticised, as there are other costs attached to the use of a resource. For this reason, travel costs will, in general, underestimate the true real price (cost) of using the resource. This may not be important, however, in estimating the benefits or consumer surplus attached to the resource. The important consideration is the shape or slope of the demand curve. If the shape of the actual demand curve can be estimated reasonably well through the use of transportation costs, then these will allow a good estimate of the benefits associated with the resource. For the same reason, if the travel itself were a recreational objective, it would not seriously affect the estimate of benefits if the shape of the demand curve remained the same.

Methods other than transportation costs have been used to estimate the demand for a recreational resource, but most of them appear to suffer from even more drawbacks than the use of transportation costs. One way of overcoming this problem is to use sensitivity analysis. Several plausable alternative values of B_o can be substituted into (6 - 1) to yield estimates of total benefit which can be compared with the costs of improving environmental quality(1).

3. A continuum of environmental quality

The special case of two possible environmental qualities is considered above. More generally, a continuum of environmental qualities is possible, and there is also a continuum of demand curves associated with these environmental qualities. Hence, a change in environmental quality will cause a shift in the demand curve. If this shift can be estimated, then the value of the change in environmental quality can be determined. (Diagram 5). This problem was tackled by Stevens(2) in estimating the demand for fishing in Yaquina Bay, Oregon. He first estimated demand equations for specific types of fishing through use of the Clawson-Knetsch methods. He regressed a fishing participation rate variable on an income variable and a variable indicating the daily fishing costs

1) For an example of this approach, see Krutilla and Cicchetti (1972) "Evaluating Benefits of Environmental Resources with Special Application to the Hells Canyon", National Resources Journal, 12(1), pp. 1-29.
2) Stevens, J. (1966) "Recreational Benefits from Water Pollution Control", Water Resources Research, 2.

per angler. The latter variable essentially represents the travel costs of the angler.

These estimated demand equations were taken to hold for existing environmental qualities. A change in environmental quality was assumed to result in an altered level of fishing success. To determine how the level of success altered the demand function, anglers in each distance-income category were asked how many days they would have fished if each of a series of six hypothetical success levels had existed. The results allowed a set of regressions of the indicated number of days fished on the hypothetical success level and the number of days actually fished.

From these regressions success elasticities were estimated. Stevens notes that there is probably some bias in these estimates because of the hypothetical nature of the questions used to determine the elasticities. Nevertheless, being the only estimates available, these results were used to determine demand curve shifts.

Once estimates for two demand curves, each associated with a different level of environmental quality, are available, the benefits of the higher quality can be estimated by the shaded area of Diagram 5. (It should be noted that the original procedure for estimating this benefit, as given by Stevens, is incorrect.) Table 7 summarises the procedure undertaken by Stevens.

4. <u>Non-transportation cost models</u>

A simpler approach for estimating the demand as a function of water quality is through questionnaire interviews. In its study(1) to estimate the benefits of water pollution control on the Lohja district, the Finnish National Board of Water asked people how often they would have gone fishing if the water quality in the given area had been high. From a comparison of these responses with existing visitation rates (at existing water qualities), the increase in the number of anglers and in the number of fishing trips per angler could be estimated, thus allowing an estimate of the benefits to be derived from improved water quality.

An alternative method of relating demand to environmental quality was employed by Deportes(2) in his study of Lake d'Annecy. Though an examination of time series, adjusted(3) tourist visitation data and water quality, he was able to find a decline in the volume

1) The National Board of Water, Finland, "The Benefits and Costs of Water Pollution Control in Lohja District".

2) Deportes, J.P. (1972) <u>Problemes économiques liés à l'assainissement du lac d'Annecy</u>, Institut National de la Recherche Agronomique, Station d'hydrobiologie lacustre, Thonon, France.

3) The number of visits was adjusted by making it relative to the size of the population of potential visitors.

Table 7

ESTIMATION OF THE BENEFITS OF INCREASED WATER QUALITY TO SALMON FISHING, YAQUINA BAY, OREGON

A. Estimation of demand for salmon fishing as a function of travel costs and income.

Specification of demand models:

$Q_{ij} = f(P_{ij}, D_{ij}, I_{ij})$

where Q_{ij} = number of angler days per 10,000 population during a 12 month period in 1963-64, for family income group i and geographic area j; P_{ij} = variable transfer costs per participant day, including an inputed round trip travel cost of 6 cents per mile, and costs of meals, lodging, and marina expenditures;(1) D_{ij} = round trip distance from residence of angler to Yaquina Bay; and I_{ij} = yearly income of angler.

The final estimated demand curve for salmon fishing from regression analysis:

$\ln Q = 5.59954 - 0.69690P + 0.53153I - 0.02214I^2$.

All of the co-efficients were significantly different from zero at the 0.01 level. The result allows the estimation of demand for salmon fishing as a whole. If income is held constant, the estimated number of angler days for alternative price increases (P_{ij}) is given below:

Price Increase per Angler Day ($)	0	0.50	1.00	1.50	2.00	3.00	4.00
Number of Angler Days Spent Salmon Fishing	10,236	7,204	5,075	3,589	2,558	1,275	564

B. Demand for salmon fishing as a function of environmental quality.

Assumption: The absence of steps to control water pollution will lead to a fifty per cent reduction in angling success.

Time series and questionnaires indicated that there was a success elasticity of between one and 0.6 for salmon fishing; i.e. for every decrease of 10 per cent in the probability of success, the reduction in fishing would be from 6 - 10 per cent.

A median success elasticity of 8 per cent can be applied to the Q_{ij}, and the model in this Table is refitted to the new Q_{ij} thus leading to a second estimate of angler days for given price increases similar to the results in Table 7.

C. Calculating the yearly benefits.

The results of the two tables can be plotted to yield two demand curves, D_0 and D_1, as in Diagram 6. D_0 represents the demand for salmon fishing at current water quality levels, and it is plotted from the data in Table 7. D_1 represents the demand for salmon fishing at reduced water quality levels.

As the current daily price of salmon fishing is at the x - axis, the total yearly benefit of the higher water quality is the difference in the consumer surpluses or the shaded area between the two demand curves D_0 and D_1.

D. Calculation of yearly benefits.

The annual benefits can be converted to total benefits as they were in the case of the Kerr Reservoir. See Table 5.

1) Stevens included the gross expenditures for meals and lodging; as indicated in the text, only the additional costs for these items can properly be considered. As the difference between the two definitions is, however, constant for all P_{ij}, this difference does not affect the regression results.

Diagram 6

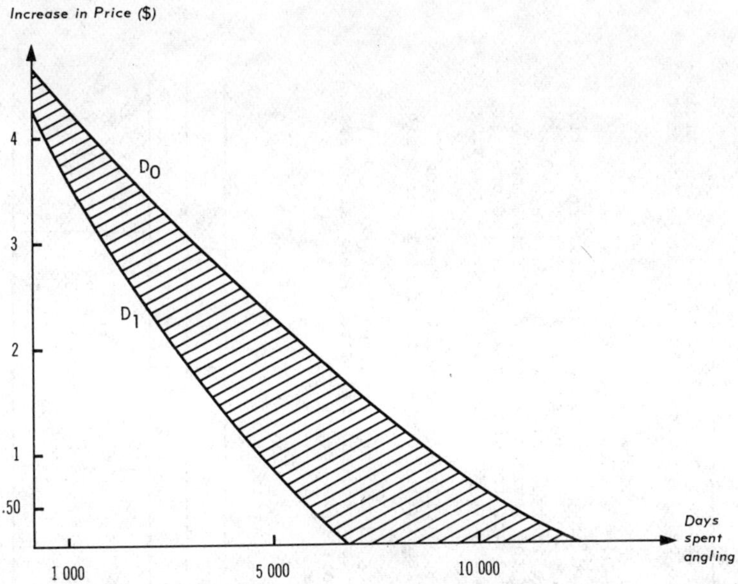

of tourist activity accompanying that in water quality. Where similar time series data exist, it is conceivable that they could be used to construct demand curves as functions of water quality.

Yet another approach to this problem is that of Davidson, Adams and Seneca(1). These authors first estimate the degree of participation in a specific activity (swimming, boating or fishing) by regressing a 0 - 1 participation variable on a set of variables which reflect socio-economic variables, as well as availability and quality of water. The data are all obtained from a national survey. Only for boating and fishing are the water quality and availability variables significant.

In applying their method to a specific area (the Delaware River Basin), the authors use the resulting regression equations to calculate current participation rates for boating and fishing for the particular population under study. The same regression equations are used to calculate the participation rates if the water facilities in the Delaware were improved. The two sets of results yield an estimate of the increased participation because of the improved water quality. Estimates of this increase were made for a thirty-year period commencing when the benefits could be realised. As there were no specific monetary values to attach

1) Davidson, P., F.G. Adams and J. Seneca (1966) "The Social Value of Water Recreational Facilities Resulting from an Improvement in Water Quality: The Delaware Estuary", in A. Kneese and S.C. Smith, editors, Water Research, Resources for the Future Inc., Washington D.C., Johns Hopkins Press.

Table 8

THE BENEFITS OF IMPROVED WATER QUALITY IN THE DELAWARE ESTURY

A. Determining probabilities of participation in recreational activities.

Data from a survey of 1,352 were used to obtain estimates of the probability of participation as a function of several socio-economic and physical variables. As an example the best regression results for fishing gave:

$$FP = \underset{(.068)}{-.0636} \underset{(.001)}{-.005A} + \underset{(.042)}{.187Y} + \underset{(.025)}{.092S} + \underset{(.019)}{.071E} + \underset{(.040)}{.174R} + \underset{(.070)}{.235W} + \underset{(.027)}{.049C}$$

The standard errors of the co-efficients are presented in parentheses below their respective co-efficients. FP is a 0-1 variable denoting presence or absence of participation; A is the age of the individual; Y represents the income of the individual; S is a 0-1 variable representing the sex of the respondent; E represents the level of educational attainment of the individual; R is a 0-1 variable indicating the race of the individual; W is the area of available sport-fishing water per capita by state in 1960; and C is a 0-1 variable denoting the presence or absence of coastal area in the region.

For those who fish, the average number of days spent fishing was 4.12. Regression analyses indicate that the frequency of days spent fishing is independent of socio-economic variables.

The above two results are applied to the existing (1960) and expected (1975 and 1990) population distribution by age, sex and race in the eleven county Delaware River Valley area to obtain estimates of yearly participation. These are given below:

FISHING (PROBABILITIES OF PARTICIPATION)

Age	1960		1975		1990	
	White	Non-white	White	Non-white	White	Non-white
18-24	.4050	.1815	.4265	.2164	.4613	.2342
25-34	.4053	.1565	.4395	.1870	.4837	.2268
35-44	.3785	.0961	.4285	.1523	.4845	.1907
48-54	.3302	.0309	.3897	.1020	.4486	.1382
55-64	.2152	0	.2988	.0076	.3548	.0629
65+	.0949	0	.1727	0	.2122	0
Participants	895,458	43,217	1,207,520	87,894	1,647,184	157,530
Total participants	938,675		1,295,414		1,804,714	
Days/part	4.12		4.12		4.12	
Total days participation	3,867,341		5,337,106		7,435,422	

If the water quality of the Delaware Valley were improved so that fishing became possible there, the W variable in the above regression would change as follows:

	1960	1975	1990
W without Delaware	.0035	.0029	.0024
W with Delaware	.0209	.0176	.0142

This new value for available per capita fishing water acreage is applied to the above regression results to construct a new table of participation probabilities, yielding a new estimate of the total number of participation days.

Those who participated in fishing were asked if they were satisfied with their frequency of participation. This yes-no (0-1) variable was then regressed upon variables reflecting available water facilities and water quality. The optimum regression results are:

$$PFP = \underset{(.0264)}{.2959} - \underset{(.0083)}{.0213}FPS$$

where PFP is the 0-1 variable reflecting satisfaction with the current level of participation, and FPS is a discrete-valued variable indicating the quality of fishing facilities as rated by experts.

In each case (with and without water quality improvements) the present value of total activity days in units of calendar days is calculated through use of the formula:

$$TAD_f = \frac{f_1}{(1+r)} + \frac{f_2}{(1+r)^2} + -- + \frac{fn}{(1+r)^n}$$

where f_i represents the total fishing activity days for year i and r is an appropriate rate of discount (taken to be five per cent in this study). The difference in the two calculations was then calculated as a measure of the increased number of days to be spent fishing as a result of improving the water quality of the Delaware.

C. Estimating the monetary benefit of improving the Delaware Estuary.

The above exercise was carried out for swimming and boating, as well as for fishing, but the regression results for swimming indicated no significant increase in that activity as a result of increased water acreage or quality. The present value in days (1965-1990) for boating and fishing were found to be as follows:

	Without Delaware	With Delaware	Differential
Boating	50,208,471	58,974,571	8,766,100
Fishing	81,611,727	82,492,451	880,724

It was decided to present the monetary value of this increased participation by applying a series of arbitrary dollar values (from \$1 to \$5) of a day's activity. These results could then be compared with the costs of improving water quality.

Source: Davidson, Paul, F. Gerard Adams and Joseph Seneca, "The Social Value of Water Recreational Facilities Resulting from an Improvement in Water Quality: The Delaware Estuary", in Water Research, Allen V. Kneese and Stephen C. Smith, editors, Johns Hopkins Press, Baltimore.

to the demand for participation, the authors applied several sets of arbitrary monetary values to an individual's participation, always relating these marginal values to a base year. A decision-maker thus confronted with the set of estimates could compare them with the costs of water improvement. A summary of this method is given in Table 8.

5. The problem of competing resources

Because the above approach considers the collectivity of water qualities over a wide area, it incorporates the problem of alternative and competing resources. A similar approach would specifically regress demand on alternative resources and activities; we shall return to this approach later, but first we examine the treatment of this problem by Cicchetti, Fisher and Smith(1) in an analysis of Mineral King in California.

The authors estimated the marginal value of a proposed ski resort at Mineral King. They specify a model of a demand system which treats all of the alternative sites. The approach is summarised in Table 9. They construct a set of linear demand equations in which demand is a linear function of travel costs estimated. Using data of visit rates by county of origin to each of the alternative sites, the authors calculate the number of visitors per thousand population for each of the fifty-eight counties of California. The travel costs to a site are calculated by multiplying the distance from the site to the population centre of each county by some fixed unit cost (per mile). The demand (visitation rate) for each zone and site is regressed on the appropriate transportation costs to other areas but subject to the restriction that the estimated co-efficients in the system obey a condition of symmetry(2). A second specification of the model eliminated those price variables whose co-efficients were negative, but not significant, because the development of an additional site does not, in general, stimulate an increase in demand in the other sites. The result allows an estimate of the demand for each area.

---

1) Cicchetti, C.J. and A.C. Fisher, "Economic Values on Environmental Resources: Some Comments by Economists on the Supreme Court Decision in the Mineral King Case", unpublished manuscript; Cicchetti, C.J., A.C. Fisher and V.K. Smith (1972) "An Econometric Model for the Valuation of Environmental Resources", presented at Econometric Society Meetings, Toronto, December 27-30, 1972; and Fisher, A.C. and J.V. Krutilla, "Mineral King Valley: Demand Theory and Resource Valuation", Chapter VII, The Economics of Environmental Resource Uses: Theory and Applications.

2) This symmetry condition is the usual one for substitution effects in consumer theory and is necessary for the solution of the system. Subsequent analyses without this condition did not allow a rejection of a null hypothesis of symmetry.

Table 9

ESTIMATING THE BENEFIT OF AN ADDITIONAL SKI RESORT AT MINERAL KING, CALIFORNIA

A. Estimating the demand for ski recreation sites in California.

Six major ski areas are considered in the analysis. These areas are at minimum potential partial substitutes for one another. Some smaller California ski sites are omitted because they are too small, or not sufficiently similar to or are too distant from the sites under study. In 1969 U.S. Forest Service survey presented data on the number of visits to ski areas by the county of origin of motor vehicles travelling to each area. Applying the average number of visitors per automobile, visitation rates (visitors per 1,000 population) for each site were calculated for each of the 58 counties of California. Travel costs between each county and site were assumed to be the shortest road distance between the site and the population centre of gravity in the county, multiplied by a cost per mile. Three alternative costs per mile ($.017, $.044 and $.104) were used in the analysis.

Regression analyses of visit rates on travel costs to each of the six sites indicated that all sites are not substitutable with one another; the northern-most site does not appear to be a substitute for the southern-most site, and vice versa. A set of regression equations of visitor rates per site or travel costs to all sites was then set up subject to two conditions:

(i) statistically insignificant negative estimated price co-efficients for other sites were eliminated and

(ii) the system of equations required that the price co-efficients by symmetrical; that is, the co-efficient of the price for site A in the regression of attendance at site B is equal to the co-efficient of the price for site B in the regression of attendance at site A.

This latter hypothesis was not rejected by subsequent statistical analyses, and the former was justified on the grounds that complementarity in the sites' services is unlikely, that is, a decrease in the use of one facility is, in general, unlikely to stimulate decreased participation at another site.

Statistically, the problem is one of estimating the regression co-efficients of a set of six regression equations subject to linear constraints upon the estimated co-efficients(1).

Site	Intercept	P_1	P_2	P_3	P_4	P_5	P_6
Tahoe	14.39(2)	-2.36(2)	.25(1)	.93	.45	--	--
June Mammoth	60.59(2)	.25	-6.23(1)	1.12	--	--	.96
Dodge Ridge - Mt. Reba	4.59	.93	1.12	-2.65	--	.48	--
Shasta - Lassen	1.05	.45(2)	--	--	-.35(2)	--	--
Badger Pass	4.29(2)	--	--	.48(2)	--	-.72(2)	--
China Peak	-1.57	--	.96(1)	--	--	--	-.77(1)

1) Significantly different from zero at the .075 level.
2) Significantly different from zero at the .05 level.

B. Estimation of benefits for the new site: Mineral King.

The total benefit of the six existing ski resorts can be estimated by integrating under the system of estimated regression curves using three different mileage costs, each of which may be plausible under certain assumptions and the ensemble of which provides a sensitivity analysis.

The new site, Mineral King, is adjacent to and very similar to an existing site; hence for analytical purposes it can be identified with the existing site. The total expected demand for this combined site can be taken as the demand for the two sites, the old and the new. This estimated demand is greater than the combined capacity of the two sites. Utilising this combined capacity constraint and recalculating the areas under the estimated regression curves yields an estimate of the consumer surplus of the ski resorts with the new site. The increase in consumer surplus over the previous estimate provides an estimate of the consumer surplus of the new site. This change in consumer surplus is found for several years and its present value is calculated. The final estimates of the present value of benefits are dependent upon the assumed mileage costs as indicated below:

Travel cost/mile	Estimated benefits
$.017	$ 3.84×10^6
$.044	$ 10.10×10^6
$.104	$ 23.84×10^6

1) Byron, R.P. (1970), "The Restricted Aitken Estimation of Sets of Demand Relations", Econometrics, 38.

Source: Cicchetti, C.J. and A.C. Fisher, "Economic Values of Environmental Resources: Some Comments by Economists on the Supreme Court Decision in the Mineral King Case", unpublished manuscript; Cicchetti, C.J., A.C. Fisher and V.K. Smith (1972) "An Econometric Model for the Valuation of Environmental Resources", presented at Econometric Society Meetings, Toronto, December 27-30, 1972; and Fisher, A.C. and J.V. Krutilla, "Mineral King Valley: Demand Theory and Resource Valuation", Chapter VII, The Economics of Environmental Resource Uses: Theory and Applications.

Total consumer surplus of the ski areas can be estimated by integrating under the resulting system of demand equations. The objective is the estimation of the change in consumer surplus from the introduction of the new facility. The qualities of the new facility are very similar to those of an existing facility, hence the total demand for the combined facilities can be estimated. The estimated demand is greater than the capacity, and the increase in consumer surplus can be estimated to be the trapezoidal area BCFE in Diagram 7. The solution of the system of demand equations along with the knowledge of the site capacities and prices allows an estimate of the increased consumer surplus. There are several assumptions inherent in the estimate: the demand function is assumed to be linear; visitors from all zones are assumed to have equal probability of access under the constraint of limited capacity; and potential changes in consumer surplus due to changes in the consumer expenditure pattern are ignored.

This method may not be suitable for all studies. First of all, it is dependent on the existence of demand data for all alternative sites. It may be particularly difficult to estimate the demand for the site under study if there is no comparable substitute. Finally, in many cases there may be partial substitutes and alternatives which merit special treatment. In this case it is necessary to use an approach similar to that of Davidson, Seneca and Adams. This is illustrated with a hypothetical example which is summarised in Table 10. We assume the existence of a data bank similar to that used in the above study. The objective of this exercise will be

Diagram 7

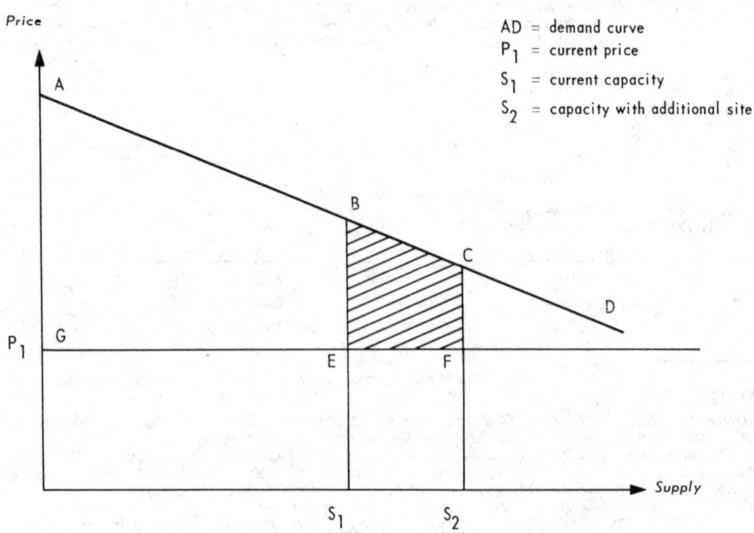

to estimate the benefit of the development of a new swimming area through a water pollution control plan. The first step would be the regression of demand for similar resources in terms of a 0 - 1 (use - non-use) variable or through use of a logistic transform. For statistical reasons, we prefer the latter approach if the available data permit.

Extensive sample survey data have to be collected giving information on socio-economic variables, water quality and availability variables and participation levels for alternative activities. The ensemble of this information could be used to construct cells for each given level of each variable. The logistic transformation can be applied to each cell by defining:

$$Z = \log \left(\frac{R+\frac{1}{2}}{n-R+\frac{1}{2}} \right)$$

where n is the total number of individuals in a cell, and R is the number which participates in the activity studied - swimming, for example. This transformed variable can then be regressed on the others. (Alternatively, but not preferably, the 0 - 1 participation variable could be utilised as a dependent variable, if necessary.) The sample survey should be designed so that the number of observations in non-empty cells does not differ by too large a factor(1).

This transformed variable and the average level of participation among those participating in each cell are regressed upon the other variables in the cell. The remainder of the analysis is similar to the Adams and Seneca approach and is summarised in Table 10.

The desirability of this type of approach is that it allows the incorporation of socio-economic and physical factors in the determination of the demand. The effects of alternative sites or even activities can be incorporated into the analysis. The analysis could be even further extended by considering a system of regressions (cf Table 9) with one regression for each of several alternative sites or activities. The resulting analysis would parallel that outlined in Table 9.

6. The choice of approach

This section introduces the Clawson-Knetsch method with various degrees of refinement and the tool of sensitivity analysis which can be applied to these and other methods. The choice of approach is dependent upon the purpose of the estimate, as well as upon the requisites of those receiving the results. Where there exists doubt about the estimates of demand curves, sensitivity analysis should be endorsed, as it provides a series of estimates under several alternative assumptions. The policy-maker is then more

1) A weighted regression analysis could be used, if necessary, to overcome this problem.

Table 10

A GENERAL APPROACH TOWARDS THE ESTIMATION OF THE RECREATIONAL VALUE OF A NEW SWIMMING RESOURCE

A. Conduct an extensive sample survey.

For each individual sampled, determine:

level of participation in swimming,

level of participation in substitute activities (fishing, tennis, etc.),

age,

sex,

income,

availability of suitable swimming resources;

quantity available,

proximity of available resources,

quality of available resources.

Factors considered unimportant for recreational demand in any area under study should be deleted.

B. Construction of cells.

For each variable, categories need to be defined, and, for numerical variables, the midpoint of each category can be used to identify that category. Every combination of categories will yield a cell. Example: males over sixty-five years with family income in the highest quartile and with a per capita area of 1-2 hectares of swimmable water within 10 kilometers of residence. Cells with very few individuals can be ignored. Within each cell i there are n_i persons of whom R_i participate in the activity studied, such as swimming. The logistic transformation is applied to these parameters, and

$$Z_i = \log \left(\frac{R_i + \frac{1}{2}}{n_i - R_i + \frac{1}{2}} \right)$$

is defined for each cell.

C. Regression analyses.

The transformed variable is then regressed on the other variables. Among those who participate, the mean level (average number of participation days) in each cell is then regressed upon the other variables. The results of these two regressions allow an estimate to be made of the level of participation in an activity, such as swimming, as a function of socio-economic and resource availability data. (cf Table 7)

D. Determining the benefit of an additional resource.

The results of the above regressions can then be applied to the population and problem under study in two cases: the existing situation and the situation anticipated after an environmental change. The difference between these two estimates will then give an annual estimate of the benefit or damage of the environmental change in terms of activity days. This difference can be calculated for several years, and the present value of the several years of activity days can be calculated as indicated in Table 7.

It remains to attach a monetary value to this physical estimate. Perhaps the best approach is the use of a range of reasonable values for per day of person activity in a sensitivity analysis. (cf Table 8) aware of the dependence of an estimate upon underlying assumptions, and his knowledge of these assumptions may influence his choice or use of an estimate.

It is recognised that, in some cases, a simple estimate is necessary, and a modified Clawson-Knetsch approach, which is refined to the extent necessary, is appropriate. The degree of refinement is again highly dependent upon the purpose of the estimate and upon the information available, but it appears as if some correction needs to be made for competing resources where these result in price (cost) changes. Additional refinements, which might be made, include the stratification of population by socio-economic and participation variables in order to obtain a more precise estimate of demand. Where the use of a resource is specialised or the population making use of the resource is limited, this additional degree of refinement is necessary. The other set of refinements are generally undertaken to improve the monetary value to be attached to the demand. The most important component in the demand for recreation at a specific resource is probably the transportation costs, over and above those which would be incurred in a visit to a similar resource. In some cases, this component of damage (benefits) unsupplemented is sufficient. It provides an underestimate of benefits, and if these are greater than the costs of undertaking the project, this estimate suggests the same alternative as would be suggested in the presence of complete information.

In other cases, the demand estimates given by the Clawson-Knetsch methods need be supplemented. Some of these demands are reflected in the increased value of property around resources. Techniques used to discuss this demand are presented in the following section.

D. MARKET REACTIONS

1. Introduction and framework

In the previous section, we saw that it is sometimes possible to draw conclusions about the monetary damage from a knowledge of the individual responses to an environmental change. It is natural to expect that individual reactions to environmental changes will have effects on the markets for private goods and in particular, on prices. In this section we propose to investigate the possibility of estimating monetary damage (or benefits) caused by a change in the environment, from a knowledge of changes in price.

The price of a private good is determined by demand and supply conditions. A proper analysis necessitates the introduction of the supply considerations. In some cases reactions to an environmental change will have no effect on prices. This is the case, for example, when a private good is produced at constant cost, and price is completely determined by supply. The simplest case to analyse is the other extreme, namely when the supply is completely inelastic, or fixed. In this case, changes in the price will be determined by demand; if the demand were to change as a result of changes in the physical environment, the monetary evaluation of these environmental changes from information on the price changes may be possible.

Land is a good in fixed supply, and we will concentrate on land for the rest of this section. If the environmental quality changes within a given area, some people may move away. Such migration leads to shifts in the demand curves for land in various districts and as a result to changes in land rents. It is possible to observe these changes in relative land rents, and they represent the capitalised value of the change in the environment. This latter statement must now be proved.

Let us consider an individual and his decision on where to buy a piece of land. We assume that he can choose between two areas which are identical except for environmental quality. He will obviously buy land in the area with the highest quality if his willingness-to-pay for the improved quality is greater than the price differential. If all individuals had identical preferences, the price differential in equilibrium would be equal to each individual's willingness-to-pay for the improved quality. If the price differential were greater, people would demand less land in the high quality area and the price differential would fall consequently. If the price differential were smaller, more people would demand land in the high quality district, causing an increase in the differential. The total value of the difference in environmental quality between the two areas is thus equal to the increased value of the high quality land compared with similar land in the low quality area.

If it is possible to investigate the relationship between environmental quality and price differentials on land, it would thus be possible to estimate the willingness-to-pay for quality improvements. But this brief discussion is based on several explicit and implicit assumptions. First of all, we have assumed that all individuals have the same preferences, which is a very strong assumption. Fortunately, the assumption can be weakened somewhat. It is sufficient to assume that different individuals' willingness-to-pay for an improvement in the environment is the same, and it is thus not necessary to assume completely identical preferences. If different individuals have different willingness-to-pay for quality improvements, the analysis becomes much more complicated. In this case, the price differential will correspond to the willingness-to-pay of the marginal individual in the high quality area and the remainder of the people in that area will earn a surplus because their willingness-to-pay exceeds the price differential. The appendix on pollution and land values provides a more rigorous development of this argument.

Most of the existing models which examine property price effects ignore the market situation. This omission can lead to serious distortions. There must be relatively unrestricted mobility, the prospective buyer must be able to perceive the environmental quality, and a majority of people must value the changes in environmental quality in a similar manner. In addition not all property must be affected; prospective purchasers must have a choice. If the environmental quality were to decrease uniformly in all neighbourhoods, property prices would be insensitive to the change.

2. Interpretation of estimates

There are many theoretical problems associated with the use of changes in property values for the assessment of damage. One practical problem is that of double counting. A change in property values reflects public perception of losses to health, materials, vegetation, and if estimates of those losses already exist, it would be incorrect to add property value changes to them. However, included in property value estimates there are probably some aspects of damage, particularly amenity losses, which are not covered in the other loss categories. For this reason, if the other categories are included in a total estimate of damage, it is the residual loss over and above the direct estimates of financial losses which should be included in a total estimate of damage. No study to date appears to have used this approach, but there are probably difficulties in implementing it because of the fact that the losses estimated through property value models may reflect a different perception of loss than that underlying other types of estimates. As an example, in the national estimates of air pollution damage made by the RECAT Committee, financial losses associated with health effects are greater than the estimated losses in residential property values. Hence, the estimated reduced environmental quality is, in fact, positive. This estimate of positive national benefit from air pollution certainly may reflect methodological deficiencies or incompatibilities among methods, thus preventing their combination.

Another interpretation of this result could be due to a difference in the public perception of damage from pollution in the vicinity of their property. At usual ambient levels, most air pollution is imperceptible to ordinary citizens. Only particulates, some strong smelling organic chemicals, and oxidants at high levels are perceptible. The immediately noticeable effects to the house-dweller are soiling in the case of particulates, the disamenity of stench in the case of some organic chemicals, and the irritation of eyes in the case of oxidants For this reason, depending upon the pollution problem present, changes in property value may only reflect certain disamenities (soiling, stench and/or eye irritation) rather than the ensemble of damage due to air pollution. In fact, working under this assumption, a Dutch team(1) felt comfortable in treating damage as estimated from property value models as a reflection of the above-mentioned amenity losses which could be added to estimates of financial losses.

3. Approaches for estimating damage

Three general methods have been used to estimate the effects of noise or pollution on property values: a direct approach, the cross-sectional approach, and the times series approach. Direct approaches have been used principally in assessing the damage due to noise. Some airports in the United States have paid residents 'fly-over' easements(2) to compensate for the noise. The value of the easements varies considerably between airports, and the determination of the easements is not clear. The threat of eminent domain may have forced many persons to accept them. For these reasons they are not considered a particularly good estimate of damage suffered by noise. Litigation results provide another source of estimates of noise damage to property, but these results exist only where there is sufficient interest and financial backing to take a risk in expending money on a court case. Moreover, settlements tend to reflect this risk, as well as lawyers' fees and, hence, may be overestimates.

The times series approach is typified by a study undertaken by Ridker(3). He compared the average sale prices of residential units in a study area and in an adjacent control area over a nine-year period. At the midpoint of the study period a new firm moved into the study area and began emitting noxious odours which led to many complaints. The results of this study are difficult to interpret: there was a divergence in property values between the two areas after the firm moved into the study area, but there were price divergences for two years before the relocation of the firm. The general difficulty with this approach is that several variables

1) Jansen, *et al*, *op.cit*.

2) i.e. residents receive compensation for the use of their air rights.

3) Ridker, *op.cit*.

in addition to environmental quality influence house prices and these variables and/or the public appreciation of them do not remain constant over time. In addition, the effects of these changes must be separated from direct environmental effects if the time series approach is to be successful.

The cross-sectional approach is the most common, and several studies have made use of it. In its general form this approach takes a measure of housing value as a dependent variable for each of several defined regions. This dependent variable is then regressed on a series of independent variables, which are believed to influence house values. A variable(1) reflecting pollution or noise is included among the independent variables. Several different dependent variables have been used including the median values of single-family dwellings, the median gross rent, the median contract rent, sale prices, property taxes, monthly rent/unit area of land, and the logarithms of these variables. Results using these several variables suggest that property taxes are not a good independent variable and that the logarithms of variables reflecting house value may be preferable to the original untransformed variables.

In general, there are two types of methodological problems encountered in this approach: mis-specification and multicollinearity. Mis-specification is generally due to the omission of some relevant variables which are related to the dependent variable and are correlated with the pollution or noise variable. Depending upon the relationship of the omitted variables with the pollution or noise variable and with the dependent variable, the result can be attributed to a positive or negative effect on the pollution or noise variable. The best way to overcome this problem is by the inclusion of all variables which could be potentially associated with housing values.

The inclusion of so many independent variables can itself lead to problems. The number of potential independent variables may be so large as to create computational problems and problems of multicollinearity. The best way of solving the former problem is the use of stepwise regression procedures, which reduce the set of independent variables. One is forced to use a reduced set of independent variables, however, if the number of observations is small. In this type of analysis multicollinearity presents a problem only with regard to the pollution or noise variable: it is conceivable that the inclusion of an independent variable which is highly correlated with the pollution or noise variable could

1) A parameter which reflects the perception of future pollution or noise levels, rather than existing or historical levels may be more important in influencing property values.

detract from the explained influence of the latter type of variable, (or at least increase the variance of its regression co-efficient) even though the latter may indeed be the explanatory variable. This type of problem may be avoided by the inclusion of several types of data sets with the aim of finding some situations where these variables are not highly associated. In such cases supplementary analyses and studies can also prove helpful. These generally involve specially designed studies whose objective is to discriminate between the effects of pollution and other variables with which it is associated.

E. LITIGATION RESULTS

There are several examples of litigation in which damages have been recovered because of changes in environmental quality, and some attempts have been made to use these in the estimation of damage. In Japan(1), for example, victims of mercury, arsenic, and cadmium poisoning and of SO_2 pollution have recovered damages. The amounts received for fatalities have increased substantially over time, and far larger amounts are recovered, for example, for pollution-related deaths than for automobile accident deaths. The latter could be due to some assumption of self-liability in an accident case whereas the victim is completely innocent in a pollution case. The variability in awards over time suggests that it may be difficult at present to attach a meaningful value to environmental damage. If the amounts awarded stabilize in the future, then these values could aid in making future damage estimates. The basis for the recoveries must be clear: is loss of amenity included in the award or are only financial losses compensated?

The present difference in the size of awards between environmental and non-environmental suits makes it difficult to use the latter to estimate environmental damage. The size of award also varies considerably when suits are similar. The Programmes Analysis Unit(2) of the United Kingdom was discouraged because recoveries varied greatly between similar non-environment cases. This is probably due to the fact that awards also reflect risks, courtroom strategies, and lawyers' fees, all of which vary according to the specific case involved and all of which must be

---

1) Morishima, A. (1975) "Damage Functions from the Legal Perspective: Japanese Legal Experiences in Pollution Damage", in Environmental Damage Costs, OECD, Paris.

2) Programmes Analysis Unit, op.cit.

considered in adapting damages recovered to estimates of environmental damage.

Other examples of highly variable awards can be found in the results of cases brought against American airports because of noise damage. Studies of property values may provide better estimates of this damage at present. If, however, such cases become sufficiently numerous, there may be some stabilization of awards, thus allowing reasonably accurate estimates of damage. School districts, Los Angeles for example, have sued because of noise annoyance and disturbance, but the damages demanded or awarded are usually related to the costs of relocating or insulating the school against noise. In such cases, these latter costs may provide better estimates than court awards on the assumption that these relocation or insulation costs are less than damage costs of noise.

Although the legal process provides a good tool for administering social justice, the damages awarded may not be defined in the same way as the damage concept incorporated in monetary estimates of damage. Legal compensation is awarded in lump sums and there may be no indication of how such a sum has been derived or what it represents. Where its definition differs from the estimate of damage defined above, the estimate derived from the legal settlement should be altered. For example, if a person whose health is impaired is awarded damages for income foregone and for costs of medical treatment, but not for psychic effects, the legal estimate should be supplemented by an estimate of psychic costs so as to arrive at a damage estimate which conforms to the definition. Where legal compensation reflects lawyers' fees, the risk one takes in bringing suit, and the time and energy thereby expended, these components would be subtracted from the total estimate.

Chapter VII

RECOMMENDATIONS AND CONCLUSIONS

A. THE CHOICE AMONG METHODS

Several methods have been and can be used for the measurement of environmental damage. "The choice of method depends upon the type of damange being estimated, the availability of information and the purpose of the study". Because concern for environmental quality is relative, necessary data and information do not always exist, and when they do exist they are usually not arranged in such a way that facilitates estimation of damage. The result is that for most damage studies undertaken to date and those to be made in the near future, it has been and will continue to be necessary to use available resources with ingenuity and in combination with a variety of methods.

Nevertheless, some methods seem to be preferable to others for estimating certain types of damage. "If the necessary data are available, the direct method appears to give the best results for most types of financial loss". In cases where the necessary data are not available, other methods such as questionnaires may be used to complement the direct method of estimation.

1. Financial losses

Defining the damage approach to be used will depend on the circumstances. General and important categories of financial loss need to be defined. The scope and purpose of the study will determine the level of detail to be used. A gross estimate of national damage may only require that a few important categories of damange be defined, whereas a damage estimate to be inserted into a benefit-cost analysis for a specific project may require a much more detailed consideration of damage.

Physical damage functions

The second step is to arrive at physical estimates of the damage within each category. Information from existing studies can be used if the levels of environmental deterioration are similar or if the effect exposure function is known. For example, if previous results indicate the percentage of a population which suffers from bronchitis because of air pollution in one area and if air pollution levels are similar in another area, it may be justified to attribute the same percentage of bronchitis cases to air pollution in the second region as in the first if the populations do not differ substantially in any way which would influence the occurrence of bronchitis. Hence the rate of pollution-induced bronchitis in one area might be extrapolated to others if the pollution levels, population exposures, and populations are similar. In some cases, more or less exact effect-exposure curves exist which allow adjustments to be made for differences in pollution level or exposure. For example, a relationship between soil acidity (due to SO_2) and growth of fir trees has been given in the literature.(1) If this relationship is accepted as reasonable, then in similar soils for any given level of acidity, estimates of growth reduction can be found.

Relationships of the latter type are rare and when they exist they are not universally accepted. The decision to use such a relationship imposes an assumption upon an estimate, and this assumption should accompany any final estimate of damage. One way to compensate for differences in opinion about the physical relationship between effect and exposure is to make estimates of damage for each of several assumptions and to present the final estimate of damage as a range, whose size is dependent upon the discordance of the underlying assumptions.

A general problem in estimating physical damage is that it is impossible to survey each category and physical area under study for all types of damage. It is often difficult enough to obtain an inventory of exposed people, materials, vegetation, etc. Hence, in many areas one is forced to make a set of assumptions which enable a damage estimate to be made on the basis of some point estimate of physical damage from another study. To go back to the bronchitis example, if the pollution-induced rate of bronchitis is known for one or a few areas, then in order to estimate this rate for an additional area where the pollution levels and/or population characteristics are different, assumptions must be made to adjust for these differences, and the worth of the estimate is dependent on that of the assumptions. No general statement can be made about the effects of such assumptions and the same assumptions may be perfectly acceptable when a gross estimate is wanted or be totally unacceptable when a refined estimate is required.

1) "Air Pollution Across Natural Boundaries: The Impact on the Environment of Sulphur in Air and Precipitation", Sweden's Case Study presented to the United Nations Conference on the Human Environment, Royal Ministry for Foreign Affairs, Royal Ministry of Agriculture, Stockholm, 1971.

Attachment of monetary values

The principal focus of this monograph has been the attachment of economic values to physical losses. "The final estimates of financial loss will be no better than the estimates of physical loss; economic approximations will only compound physical approximations". On the other hand a gross physical estimate probably does not merit exceedingly refined economic techniques. Several economic factors influence the monetary value attached to a physical damage. Of these factors, the final estimate is the most sensitive to the monetary value attached to the damaged product, material, etc. In some cases this value can be found easily as, for example, the wholesale prices of an agricultural product. In other cases the choice of a value is more difficult. For example, if air pollution damage to steel bridges in a wide area were being considered, it would not be possible to consider each bridge separately, and a norm needs to be defined both for the amount of physical damage suffered by a bridge and for the value of this damage. The problem would be compounded if the category were broader, as would be the case if steel products in general were considered rather than specific end-products using steel, such as bridges. The optimum level of generalisation for valuation purposes cannot be specified. It depends upon the information available and upon the objectives of a particular study. In general, the broader the categories, the greater the assumptions which need to be made and the greater the uncertainty which must be attached to the final monetary estimate.

The definition of monetary damage given here assumes that individuals and firms behave optimally, and this assumption leads to a unique definition of monetary damage. For this reason the monetary values attached to the physical damage should reflect this optimum behaviour. Hence, if the physical damage to a bridge as a result of pollution can be compensated for completely by either repainting or replacing a bridge, then the smaller of the compensation costs (those incurred in repainting or in replacing the Bridge) should be applied in the estimation of the monetary damage.

Accounting problems

There are two principal accounting problems, and both relate to the treatment of time. The first is the proper attribution of damage to the time period during which environmental deterioration was responsible for the damage. Again this procedure will involve the making of several assumptions to allow some attribution of damage to a specific time period. For example, if several years of air pollution are responsible for the necessity of premature repainting of a building, some assumption must be made about how to allocate the damage among the several years during which pollution induced it. The second problem area is the choice of an appropriate discount rate. The discount rate should be the same as that prevailing in other social welfare projects. One problem in estimating these rates is that they may vary. Moreover, the final estimate can be sensitive to the chosen discount rate: an increase in the rate from five per cent to ten per cent in estimating the present value of lost earnings due to premature death gave Ridker.(1) a twenty-five per cent reduction in his estimate. If the decrease in lifetime because of environmental effects is relatively small, the estimate of loss will be relatively insensitive to the choice of an interest rate.

The remaining economic consideration in the application of monetary values is the effect of environmental quality changes on supply and hence on the prices to be attached to physical damage. An examination of existing studies indicates that decreased environmental quality reduces supplies of some products, agricultural products for instance, but that these decreases have not been sufficiently large to affect price. Hence, outside of exceptional circumstances, this consideration can probably be ignored.

2. Losses of amenity

The problems involved in estimating losses of amenity are more severe than those involved in estimating financial losses. Working with intangible factors, it is much more difficult to arrive at and verify estimates. Even greater ingenuity in the use and choice of a method is required, particularly because specific amenity losses are better suited to some methods than to others. Table 11 attempts to indicate those methods which can provide the best damage estimates for specific amenity losses.

Health effects

The most difficult types of amenity loss are those due to effects on human health: the costs of suffering, bereavement, premature death, inconvenience, etc. No method to date has satisfactorily estimated these losses in amenity, although they are probably as great as the associated financial losses. For this reason these losses cannot be ignored.

Past decisions have implicitly placed some value on such amenity losses, but these values are difficult to determine: they are often weighed against other intangible factors which involve other parameters, or show no consistent values. For example, it is

1) Ridker, op.cit.

Table 11

SUGGESTED METHODS FOR ESTIMATING TYPES OF AMENITY LOSSES

Damage Category	Loss of Amenity	Suggested Methods	Comments
Human Health	cost of suffering cost of bereavement cost of limitatations imposed upon an individual, his family and his society	litigation results	may lead to over-estimates if extra-polated to complete affected population
		in cases such as that of noise, where an individual may at least partially insulate him-self from the source of annoyance or suffering, the costs of protecting himself against the nuisance	assumes rational be-haviour
	cost of annoyance		can lead to under-estimates
		modified questionnaire approach using tradeoffs (Hoinville & Prescott-Clark)	methods need further study require significant resources and care in execution (see Table 6)
Fauna	reduced pleasure from hunting, fishing, wildlife viewing	modified Clawson-Knetsch method (utilising changes in visitation or activity participation rates)	continued refinement necessary; at present, several drawbacks (see Table 7); takes no account of future generations
		use of the cost of build-ing and maintaining alternative reserves or resources	assumes that construction costs are no larger than the benefit imperfect substitutes
		sensitivity analysis	
Flora	reduced pleasure from horticultural and forestry losses	same as for Fauna	
Natural Resources	decreased recreation benefits	modified Clawson-Knetsch	(see above and section V - D)
		in some cases, the costs of constructing substi-tutes	such substitutes are lar-gely imperfect
		sensitivity analysis	
Materials	endurance of soiled or damaged materials	costs of cleaning, re-pairing, replacing, or protecting materials	may lead to overestimates
	damage to selected aesthetic monuments and articles	use of insured values	may not exist
		costs of restoration	restored item may not have same value

well recognised that a reduction in speed limits for automobiles would reduce the risk of death from motor vehicle accidents, yet other societal concerns (appreciation of time, pleasure of driving fast, etc.) and concern for efficiency weigh against speed limit reduction. As there are many factors involved (some of which are intangible) in the decision to set a speed limit, it is extremely difficult to determine any upper bound on the implicit value placed upon a reduction in the risk of death from this decision. Other examples are those in which a decision is made on the maximum amount of medical care to be given to an individual. The sums for such care vary greatly and are not necessarily correlated with the degree of suffering and amenity loss affecting an individual. Further consideration of past decisions may yield some rough estimates of the extent to which a society values human health or life, but at present such investigations have not been sufficiently exploited to provide even rough estimates.(1)

The only other method which seems potentially useful for obtaining estimates of disamenity from health effects is the use of litigation results. The disadvantages of this method have been cited above, but it can be hoped that as the number of cases increase, the consistency in amounts awarded will improve, thus enabling better estimates of damage to be made.

At present, however, we are left without any reliable estimates of this important damage category. This is no excuse to ignore this damage; it is too important to ignore. The best solution appears to be to accompany monetary estimates of other types of damage by some physical indicator of health effects. Those who must make a decision can then weigh this information before making a decision.

Noise nuisance

In cases such as that of noise, where some action can be taken to protect against the disamenity, the cost of protection can be used to provide some estimate of the damage. With noise, for example, there is some knowledge of the relationship between noise level in NNI (Noise and Number Index) and the percentage of individuals seriously annoyed.(2)

Hence, for a given level of noise in an area, it is possible to estimate the number of people seriously annoyed. There is no straightforward way to transform this physical variable into a

1) Using actuarial and wage rate data for workers in very dangerous occupations, one study estimated the disamenities associated with death to be about three times the expected discounted future earnings. See Thaler, R. and S. Rosen (1973) "The Value of Saving a Life: Evidence from the Labor Market", Graduate School of Management, University of Rochester, Rochester, N.Y.

2) See, for example, Alexandre, A. and J-Ph. Barde (1973) Le Temps Du Bruit, Flammarion, Paris.

monetary one. It is possible to use the costs of protecting against noise through double-glazing or the use of double-windows as an estimate of annoyance.

There are many problems involved in the above measurements. First of all, the cost of protecting against noise is very high and requires a considerable outlay, straining the financial resources of some households. For this reason, one might consider looking at the beginning of the behaviour of those households whose income is above some specified level. If, for example, twenty-five per cent of the household examined instal some kind of double-window to protect against noise, this percentage could be applied to the costs of insulating all households in the noisy area, thus allowing a monetary estimate. This will probably be an under-estimate as sound-proofed housing is not a perfect substitute for a noise-free environment: the out-of-doors cannot be protected and windows cannot be opened if sound-proofing is to be maintained. Moreover, this estimate attributes no damage to those who are not sufficiently annoyed to install noise-protection devices. Yet another drawback is the fact that insulation against noise is also an energy conservation measure, hence noise nuisance may be only a partial motivation for installation.

Use of Clawson Knetsch methods

Losses of amenity from damage to fauna, flora and natural resources can probably be best evaluated through the use of case studies which attempt to estimate the demand for these resources. The methodology in this area continues to be developed, but the modified Clawson-Knetsch method appears to be the most widely accepted. Where loss of amenity can be related to a decrease in demand for a specific area or activity, the change in demand due to an environmental quality change can be estimated by methods similar to that of Stevens. An additional refinement is probably needed in most cases to account for possible substitution among areas or activities. Cicchetti et al have indicated how to account for site substitution in a specific case. General models which make additional assumptions about substitutability of areas and activites need to be developed. In some cases where an area is unique or the prices attached to its use are unaffected by alternative areas, existing models similar to the Stevens approach can be used.

Use of direct methods

To a lesser extent, it may be possible to use the costs of building or maintaining alternative facilities as an estimate of the damage. The cost of constructing or maintaining a wildlife reserve may give some indication of the value of preserving a certain species threatened by environmental degradation. The same approach could be used to value recreational areas and fishing and swimming in some cases.

The costs of restoration can sometimes be used as an estimate of damage, as in oil spills or strip mining. The use of this estimate implies, however, that the damage is at least as great as the restoration costs; otherwise, restoration would not be a rational action. Given the nature of such an estimate, its uses are limited. It could not, for example, be used in a benefit-cost analysis to justify the restoration, but such estimates could be useful in comparing the damage of alternatives. In a comparison of the damage resulting from the selection of certain fuels, for example, the costs of restoring strip-mined land and those of cleaning up oil spills can be included in estimates of the environmental damage associated with the use of coal and oil respectively.

Use of questionnaires

Direct questions which ask how much people would be willing to pay for a specific resource do not yield consistent estimates of the value of a resource but variations on this approach may prove to be profitable. The Hoinville-Prescott-Clarke experiments indicate that the "priority evaluator approach" has some potential, but further experimental results are needed before this method can be recommended. One of its drawbacks is that it requires considerable resources and time to implement, and these may not be available for many projects for which damage estimates are needed.

Market studies

Attempts can be made to estimate the losses of amenity due to soiling, noise and aesthetic disadvantages through the study of market reactions. In general, this approach is not very satisfactory, as property value models suffer from many problems, which have been outlined in chapter VI. Additionally, it is difficult to extricate the losses of amenity from the financial losses in these models when both occur. One way to overcome this attribution problem is to ignore certain financial losses which are assumed to be accounted for in property models. One study,(1) for example, in looking at the damage cost of air pollution, tried to estimate changes in property value from changes in environmental quality.

---

1) Jansen, et al, An Estimate of Damage Caused by Air Pollution in the Netherlands in 1970, English Summary, Institute for Environmental Problems, Free University of Amsterdam.

Aware of the double-counting problem if this estimate were added to financial estimates, the authors of this study decided to ignore those financial losses which are probably included in property value loss, as soiling costs and damage to housing materials.

Sensitivity analysis

A refinement which can be applied in some circumstances is related to sensitivity analysis. Different reasonable values can be attributed to input parameters of other methods or an amenity loss to itself and the impact of the alternative assumptions on the final estimate of damage can be seen. The set of final estimates can also be accompanied by the assumed values, thus allowing the policy-maker to choose the final estimate using the value which he feels is most reasonable. An example where several values were attached to the amenity loss is given in section VI.C. In that example, Davidson *et al*(1) estimated the value of improving the quality of the Delaware River Basin. In their study, the authors attached several different values to an individual's participation-day thus yielding several final estimates of the recreation value of improved water quality in the Delaware River Basin. The conclusions of this study consisted of several estimates for recreational value, each of which is accompanied by its assumed value of an individual's recreation-day.

B. THE USES OF DAMAGE ESTIMATES

Environmental damage estimates can be used for a variety of purposes: for benefit-cost analyses, to compare alternative projects or policies, in litigation, to determine national and regional priorities, and to set charges on pollution emissions. The usage of a given method is somewhat dependent upon the purpose for which the estimate is made, and the objective determines how useful and how approximate a given estimate can be.

1. Benefit-cost analyses

The benefit-cost analysis requires an estimate of the (marginal) benefits of a project or action and this value is compared with the costs involved in undertaking the project or action. Depending upon the subject of the analysis and upon the specific definitions chosen, damage estimates can be included in either side of the benefit-cost equation, as damage avoided on the benefit side or as an external cost on the cost side. The accuracy demanded of monetary damage

1) Davidson, P., F.G. Adams and J. Seneca (1966) *op cit*.

estimates will depend upon the size of the difference between the estimates of the two sides of the benefit-cost equation. If the costs are much greater than the benefits, then even a sizeable error in the estimate of either side of the equation will allow the same decision to be reached. The magnitude of the difference between the two sides of the equation is not known in advance (or the analysis would never have been undertaken), but if the resulting difference is large, substantial confidence can be placed in the result of the analysis.

If the estimates of environmental damage are known to be biased, and the direction of the biases are known, the use of the biased estimates in benefit-cost analyses may not necessarily detract from these analyses. If, for example, the damage estimates is known to be an underestimate and the side of the benefit-cost equation in which it appears is greater than the other side, little information has been lost. The benefit-cost analysis indicates the same alternative as a similar analysis with perfect estimates would indicate. Similarly, if the damage estimate is known to be an overestimate and the side of the benefit cost equation in which it occurs has the smaller value, the result is as satisfactory as it would be with perfect damage estimates.

Because known underestimates or overestimates of environmental damage may be as valuable as precise ones, it is important to know the direction of bias in a damage estimate. For this reason "if there are several components of damage in an overall monetary estimate, it is necessary to ensure that all components with significant biases are biased in the same direction". The direction of the bias of the total monetary estimate will then be known.

2. Evaluation of alternatives

When the environmental effects of alternative processes, technologies or locations are compared, the components of damage are likely to be similar. For this reason, the estimates of damage associated with the alternatives are based upon the same assumptions and they are likely to be biased in the same direction. Thus a comparison of alternatives is possible when the monetary estimates of damage associated with the alternatives are rough. The greater the similarity of the types of effects among the alternatives, the easier it is to accept rough estimates. If, for example, several locations were under study for the siting of a new industrial complex, whose future environmental effects can be estimated, and it is known that agricultural production and human health are the likeliest victims of adverse effects associated with this new complex, a comparison of the monetary damage associated with each location might facilitate a choice of site. A meangingful comparison of sites is likely to be possible, even if health damage were subject to significant bias.(1) Where there is a greater difference in the types of effects associated with alternatives, the above is less true. Hence, if two types of industries were under consideration, the emissions from one of which caused health damage and from the other agricultural damage, then any comparison of their effects could be very sensitive to estimation errors if the difference in effects were small. Often, however, alternative projects or actions produce similar types of effects.

One advantage of this use of damage estimates is that if a particular component of damage cannot be evaluated in monetary terms or can be evaluated only very poorly, a vector of damage can be presented for each alternative, some components of the vectors being expressed in monetary terms and others in physical terms. The vectors can then be compared as part of the choice process, and the use of monetary estimates has reduced the number of components in the vector.

3. Charges

Damage estimates can serve as a basis for setting charges on marginal emission to the environment. In this role as a policy instrument, the marginal global damage resulting from marginal emissions would be assessed to each polluter. This use requires estimates of the total marginal damage, and as these estimates are, in general, very rough, it may be better to use other bases for charges at present.

4. Global estimates

Several national or regional estimates of a type of damage (usually as a result of air pollution) have been made. These have given governments an estimate of the magnitude of the environmental problems they face and have helped to place these problems in a priority list along with other problems. Governments are then able to allocate their resources according to these priorities. Even within the environmental sphere, priorities can be obtained from damage cost estimates: damage by category and by type of pollution can be ordered. Thus, for example, if air pollution damage were much greater than another kind of pollution, priority can be given to air pollution abatement. Similarly, if human health damage were greater than other types of damage, national policies might be oriented towards giving priority to reducing health damage.

1) It is possible to conceive of cases where such bias could lead to an incorrect decision, but, in general, the greater the similarity of the damage components associated with the alternative, the more tolerable are inaccurate estimates.

This is an area of use where damage cost estimates need not be precise. A rough estimate can often suggest the priority which needs to be given to a particular area. If the resulting estimate is a paltry sum, even if it is inaccurate by an order of magnitude, the attention merited by the problem is much less than if an inaccurate, but large, estimate of damage were obtained.

A second use of national or regional estimates of damage is to adjust national income statistics for environmental damage in order to have a better national or regional welfare index. Current overall estimates of environmental damage are subject to much greater error and uncertainty than are national income statistics, and for that reason it is probably not wise at present to combine the two into an index.

C. GENERAL CONCLUSIONS

Use as a planning tool

"When a major intervention in the environment is contemplated, the decision should consider attempts to estimate the monetary damage associated with the intervention". Examples are hydropower development, the siting of heavy polluting industries in new areas, etc. Damage estimates can facilitate the planning process associatod with such interventions and can aid in the decisions to be taken. It is nevertheless realised that damage estimates will not be the only input into the decision-making process, and other factors may be much more important in influencing the final decision. Although perfect monetary estimates may not be available, attempts to construct them may force some additional degree of rationality on the decision-making process and will, at a minimum, provide some new information which may aid in the decision-making process.

Necessity of stating underlying assumptions

"The assumptions upon which damage estimates are based must be accessible to all persons using the final estimate". In general, these people will not be those who constructed the estimate. This fact helps to ensure that environmental damage functions are not the sole criteria for a decision, but it also necessitates a clear exposition of the assumptions underlying the estimate. When there are several reasonable alternative underlying assumptions, "it may be best to express the estimated damage as a range reflecting the effects of the alternative assumptions", rather than as a single point estimate. This will minimise the possibility that a point estimate of damage will be misinterpreted and misused.

Statement of conceptual framework

The underlying assumptions include the conceptual basis; which must also be clear to the user. In this monograph, the conceptual basis of monetary damage functions has been the individual's compensating variation, which seems to suit best economic theory and a market-oriented society.

One of the main purposes of this monograph has been the presentation of a conceptual basis. In the present treatment of monetary damage, income redistribution is not specifically considered. This is because income redistribution goals vary considerably. If these goals are known and if it is desired to use environmental policy to help implement them, "the methods presented here can be adapted to take into account any desired distribution of income".

Most promising areas for good estimates

Financial losses are, in general, easy to determine if the physical damage associated with them is known because they are defined as changes in the level of expenditure and these changes can usually be found. For this reason, types of damage which are composed principally of financial losses can, in general, be estimated quite accurately. Hence, reasonable estimates can be found for materials damage and lost agricultural productivity, as financial losses predominate in these categories of damage. Experience also indicates that acceptable estimates are often achievable for damage to recreational resources, although some further research is needed to refine these estimates when there are accompanying price changes in competing resources.

Qualifications on use

There are two major problems associated with the use of monetary damage functions:

a) the lack of knowledge about the relationship between emissions and ambient concentrations and the relationship between ambient concentrations and physical damage;

b) the problems associated with putting a monetary value on the physical effects.

Only the second problem has been covered in this monograph, but any serious use of monetary damage functions requires that efforts be made to increase understanding about the first type of problem. Lack of knowledge about the chemical, physical and ecological effects of waste discharges resulting from direct interventions in the environment is, however, a general problem in environmental policy.

Given some knowledge of the physical effects of waste discharge, there remains the problem of monetary evaluation. "For complex problems, where there are many different types of environmental effects, some of which are amenity losses, it appears impossible at present to estimate monetary value accurately". In particular, "when there are irreversible environmental changes present, unbiased estimates of the true monetary damage are not attainable".

Usefulness of imperfect estimates

"It is not always necessary to have a perfect estimate of the monetary damage". Sometimes an underestimate will be sufficient. Whenever this underestimate is greater than the cost of cleaning up the environment, a decision can be made. Moreover, even if the monetary damage estimate is below the polution abatement cost, but the nature of the physical damage which has not been included in the monetary damage is known in a qualitative way, there is a better opportunity of arriving at a rational decision. Similarly, an overestimate of damage will be sufficient if it is smaller in value than the costs of cleaning up the environment.

Imperfect damage estimates can also be used in comparing alternative actions, technologies or sites. Where poor estimates of damage components exist, the comparisons can be facilitated through use of sensitivity analysis or a variation of multi-criteria analysis in which effects are presented in vector form, as monetary estimates when available and as physical estimates when monetary estimates are not available.

Necessity for further work and co-operation

Research on the estimation of losses of amenity has been mostly carried out in one discipline; economics. An interdisciplinary approach, involving closer co-operation between natural scientists, sociologists, psychologists and economists, may indicate new and better approaches for the quantification of amenity values. In addition, this co-operation may yield better predictions of environmental changes, and it will guarantee that information about the natural environment is organised in a way that is useful for decision-making purposes.

Appendix A

MATHEMATICAL APPENDIX TO CHAPTER III

The purpose of this appendix is to give a fairly self-contained presentation of the material discussed in Chapter III.

Consider an individual whose preferences can be represented by a utility function:

$u(c, Y)$

where c is the vector of net demand of the n different goods and services, and Y is the vector of the flow of environmental services. In accordance with the conventions in III.C, we assume that the utility function is increasing in the different components of Y.

The Marshallian demand functions are defined from:

$$(1) \quad \max u(c, Y)$$

$$\text{subject to: } \sum_{i=1}^{n} p_i c_i \leqq I.$$

where c_i is the net demand for good i and p_i the corresponding price where $p_i \geq 0$.

They are denoted $c = c(p, Y, I)$.

The compensated demand functions are defined from:

$$(2) \quad \min \sum_{i=1}^{n} p_i c_i$$

subject to $u(c, Y) \geqq \bar{u}$

where $\bar{u}$ is the utility level in situation A.

The compensated demand functions are alternatively denoted by:

$c^*(p, Y, \bar{u})$ or:

$c^*(p, Y, A)$.

The expenditure function is defined by:

$$(3) \quad m(p, Y, A) = \sum_{i=1}^{n} p_i c_i^*(p, Y, A)$$

and is interpreted as the smallest income necessary to make the situation characterised by the price vector p and the environmental quality vector Y indifferent to situation A (or to make the utility equal to $\bar{u}$). With this interpretation as a background, it is clear that:

(4) $c[p, Y, m(p, Y, A)] \equiv c^*(p, Y, A)$.

If we write down the Lagrangian for (1) and put the partial derivatives with respect to c_i equal to zero, we get:

(5) $\frac{\partial u}{\partial c_i} - \alpha p_i = 0, \; i = 1, \ldots, n$

(where α is a Lagrangian multiplier) as necessary conditions which the uncompensated demand functions must satisfy. The corresponding necessary conditions for (2) are (with β as a Langrangian multiplier):

(6) $p_i - \beta \frac{\partial u}{\partial c_i} = 0, \; i = 1, \ldots, n$

If we differentiate (3) with respect to p_j, we obtain:

$$\frac{\partial m}{\partial p_j} = c^*_j + \sum_{i=1}^{n} p_i \frac{\partial c^*_i}{\partial p_j}$$

But we know that $u(c^*, Y) = \bar{u}$, and differentiating this equation with respect to p_j yields:

$$\sum_{i=1}^{n} \frac{\partial u}{\partial c^*_i} \frac{\partial c^*_i}{\partial p_j} = 0.$$

Together with (6) this yields:

$$\sum_{i=1}^{n} p_i \frac{\partial c^*_i}{\partial p_j} = 0, \text{ and so}$$

(7) $\frac{\partial m}{\partial p_j} = c^*_j(p, Y, A), \; j = 1, \ldots, n.$

The partial derivative of the expenditure function with respect to the jth good price is therefore equal to the compensated demand function for the jth good.

Let situation A be characterised by $\bar{p}$, $\bar{Y}$ and assume that the only change is in the price of good number one. Because of (7) we have:

(8) $$m(p, Y, A) = m(\bar{p}, \bar{Y}, A) + \int_{\bar{p}_1}^{p_1} c^*_1(p, Y, A)dp_1.$$

but $m(\bar{p}, \bar{Y}, A)$ is equal to I, and (8) therefore implies that the compensated variation is equal to the integral of the compensated demand function for good number one from $\bar{p}_1$ to $\bar{p}_1$, that is, the area under the compensated demand curve.

Let us now differentiate (4) with respect of p_j; We obtain:

(9) $$\frac{\partial c_i}{\partial p_j} + \frac{\partial c_i}{\partial I}\frac{\partial m}{\partial p_j} = \frac{\partial c_i^*}{\partial p_j}$$

Using (7) and rearranging terms we have:

$$\frac{\partial c_i}{\partial p_j} = \frac{\partial c^*_i}{\partial p_j} - c^*_j \frac{\partial c_i}{\partial I}$$

The left-hand side is the effect on demand of good number i resulting from a change in the price of good number j. On the right-hand side, this effect is divided into a substitution effect $\partial c^*_i/\partial p_j$ and the income effect $c^*_j(\partial c_i/\partial I)$. If the income effect is small in comparison with the substitution effect it is seen that it is possible to approximate the compensated demand function by the uncompensated demand function.

Using (7) we can rewrite (9) as:

(10) $$\frac{\partial m}{\partial p_i \partial p_j} - \frac{\partial c_i}{\partial I}\frac{\partial m}{\partial p_j} - \frac{\partial c_i}{\partial p_j} = 0$$

which is another useful formulation of the Slutsky equation.

The partial derivative:

(11) $$\delta_j = + \frac{\partial m}{\partial Y_j}$$

will be called the demand price for environmental service j.

By partially differentiating (3) we have:

$$\partial_j = \sum_{i=1}^{n} p_i \frac{\partial c_i^*}{\partial Y_j}.$$

But as $u(c^*,Y) = 0$, we also have:

$$\sum_{i=1}^{n} \frac{\partial u}{\partial c_i^*}\frac{\partial c_i^*}{\partial Y_j} + \frac{\partial u}{\partial Y_j} = 0$$

From (6) we see that this entails:

$$p_i \frac{\partial c_i^*}{\partial Y_j} = - \beta \frac{\partial u}{\partial Y_j} \text{ or:}$$

(12) $$\delta j = \frac{p_i \frac{\partial u}{\partial Y_j}}{+ \frac{\partial u}{\partial c_i}}$$

The demand price for environmental service number j is thus equal to the value of the marginal rate of substitution between the environmental service and a private good.

Assume now that the environment changes, so that the vector Y, describing environmental quality, changes from Y' to Y". Label the initial situation A. Assume that the change is a degradation of environmental quality. The monetary damage is then:

$$(13)\ CV^h = m^h(P, Y'', A) = I^h,$$

where I^h is the wealth of individual h.

The equivalent variation is defined as:

$$(14)\ EV^h = I^h - m^h(P, Y', B)$$

where B is the situation after the change.

The benefit from an improvement of environmental quality is defined as the negative of CV^h of:

$$(15)\ B^h = I^h - m^h(p, Y'', A).$$

Aggregation Over Individuals

Total monetary damage is defined as the sum of the individual damages:

$$(16)\ CV = \sum_{h=1}^{H} CV^h.$$

Assume that the environmental quality changes from Y' to Y" because a plant is constructed. The profit of the new plant is expected to be R.

If:

$CV < R$

and if lump sum transfers are possible the profit can be distributed among other individuals in such a way that no one is worse off and at least one individual is better off. In this case the plant should be constructed. If $CV > R$ no such redistribution of profits can be made, and at least one individual is hurt by the construction which therefore should not be undertaken.

If lump sum transfers are not feasible and if equity questions do not matter a comparison between CV and R is still meaningful. The project should be undertaken if $CV < R$ and should not if $CV > R$.

If, however, lump sum transfers are not feasible and if equity does matter, a direct comparison between CV and R is no longer relevant and another definition of total damage must be chosen.

Assume that the inter-personal values of the decision-maker can be represented by a social welfare function:

$$(17)\ W(u^1, \ldots, u^H).$$

Suppose we have an initial situation A in which the individuals have utility levels $\bar{u}^1, \ldots, \bar{u}^H$, respectively. Assume now that the quality of the environment changes from Y' to Y". Because of this change some individuals are better off and others are worse off. The change in utility functions can be denoted by Δu^h.

The welfare change is then:

$$\Delta W = W(\bar{u}^1 + \Delta\bar{u}^1, \ldots, \bar{u}^H + \Delta u^H)$$

According to the mean value theorem of calculus, there exist constants α^h, θ such that $\bar{p} \leq \theta \leq 1$,

$$\alpha^h = \frac{\partial W(\bar{u}^1 + \theta \Delta u^1, \ldots u^H + \theta \Delta u^H}{\partial u^h}, \quad h = 1 \ldots H$$

and

$$\Delta W = \sum_{h=1}^{H} \alpha^h \Delta u^h$$

Consider:

$$m^h(\bar{p}, Y', u^h)$$

as a function of u^h, i.e. keep $\bar{p}$ (the price vector in the initial situation) and Y' constant $m^h(\bar{p}, Y', u^h)$ is then a monotonically increasing transformation of the utility function u^h and is therefore a representation of the preferences of the individual. We can therefore assume that:

$$\Delta u^h = m^h(\bar{p}, Y', u^h) - m^h(\bar{p}, \bar{Y}, \bar{u}^h).$$

We also have:

$$m^h(\bar{P}, \bar{Y}', \bar{u}^h) = I^h, \text{ and thus:}$$

(18) $\Delta u^h = m^h(\bar{p}, Y', u^h) - I^h = EV^h.$

The welfare change is therefore:

(19) $\Delta W = \sum_{h=1}^{H} \alpha^h EV^h$

With this representation of the preferences it can be shown that α^h is the value the decision-maker puts on a marginal increment of the hth individual's nominal income.

If $\Delta W > 0$, the change is worth undertaking because it increases the welfare, while if $\Delta W < 0$, the change should not be undertaken.

Risk

Assume that the initial situation is characterised by price vector p, income I, and probability zero of contracting a respiratory disease. Then the air becomes polluted and the probability of contracting the disease increases to Pr. The problem is to analyse the appreciated monetary damage.

Assume first that Pr = 1. Then the individual knows with certainty that he will contract the disease, and the previous analysis can be applied. Let the individual's utility function be u(c, x) where x is a dummy variable, taking the value a when the individual is healthy and (b) when he is sick.

Utility maximisation gives demand functions:

c(p, I, x).

The maximum utility that can be achieved is then:

(20) $v(p, I, x) \equiv u[c(p, I, x), x]$

where $v(p, I, x)$ is the individual's utility function.

The compensating variation $\overline{CV}$ can now be defined by:

(21) $v(p, I + \overline{CV}, b) = v(p, I, a)$.

Let us now consider the case when $0 \leq Pr \leq 1$.

By assuming the axioms underlying the von Neuman Morgenstern expected utility theory, the relevant utility function for the choice between different probability distributions can be written:

(22) $u = (1 - Pr)\, v(p, I, a) + Pr\, v(p, I, b)$.

The compensating variation CV is then defined by:

(23) $(1 - Pr)\, v(p, I + CV, a) + Pr\, v(p, I + CV, b) = v(p, I, a)$

In general, it is very difficult to estimate CV but it is possible to get some information about $\overline{CV}$ (as the example in the text shows). If the consumer is risk neutral and if the marginal utility of income is the same in the two situations it is possible to show that:

$CV = Pr\, \overline{CV}$.

Assume first that the individual is risk neutral, i.e. v is the linear in income.

(23) can then be written:

$CV[(1 - Pr)va + Pr\, vb] = Pr[v(p, I, a) - v(p, I, b)]$

where $Vx = \frac{\partial v(p,I,x)}{\partial I}$

(21) can be written:

$Vb\, \overline{CV} + v(p, I, b) = v(p, I, a)$.

This yields:

$CV\{va - Pr(va - vb)\} - Pr\, vb\, \overline{CV} = 0$

If the marginal utility of income is equal in the two situations, i.e. va = vb, we have:

(24) $CV = Pr\, \overline{CV}$.

However, if v is non-linear, (24) may not be a good approximation.

Appendix B

HEALTH EFFECTS AND PREMATURE DEATH

The definition of monetary damage given in Chapter III implies that the cost of an increased probability of pollution-induced health effects or of premature death is the amount by which the individuals must be compensated for the increased risk. This concept is presented here in more detail.

Assume that in the initial situation the probability of contracting a pollution-induced disease is zero; but, due to environmental degradation, this probability increases to p_1. Assume further that the conditional probability of dying from the disease is p_2.

Consider now an individual h who is exposed to the increased risk. If he gets the disease, the total cost is given by the example in Chapter III and consists of medical expenditures ME^h, wages foregone W^h and psychic costs PC^h. In general he is not compensated completely for these costs, but assuming that he only has to pay a fraction α of the medical expenditures and that he gets sick pay so that he loses only the fraction β of his wage earnings, the cost to the rest of society is then:

$$(1-\alpha)ME^h + (1-\beta)W^h.$$

If we assume that the rest of society is risk neutral the expected cost is then:

$$p_1 \left\{ (1-\alpha)ME^h + (1-\beta)W^h \right\}$$

If the individual dies, the cost to society would consist of the net financial contributions between the individual and the society, which would have taken place if the individual had not died (neglecting the psychic costs to near relatives and friends). Assume that the individual dies at age N^h and that his life expectancy in the absence of the increased risk is $\bar{N}^h$. Let the net transfer per year from the individual to the rest of the society be NC^h. Thus the total cost to the rest of the society is:

$$\sum_{t=N^h}^{\bar{N}^h} \frac{NC^h}{(1+r)^{t-N^h}}$$

The total expected cost is thus:

$$p_1 \left\{ (1-\alpha)ME^h + (1-\beta)W^h \right\} + p_2 p_1 \sum_{t=N^h}^{\overline{N}^h} \frac{NC^h}{(1-r)^{t-N^h}} = CV_r^h$$

Where CV_r^h is the expected total cost to the rest of society resulting from exposing individual h to a greater risk. A principal component of NC^h is the taxes net of any transfers and personal consumption of publicly provided goods and services. Another component is the net savings.

Let us now look at the cost to the individual. That cost is defined as the amount necessary to compensate him for the greater risk he is exposed to. Call this CV^h. The total social cost is thus:

$$TSC = \sum_h \left\{ CV^h + CV_r^h \right\}.$$

It is clear that CV^h will depend on αME^h and βW^h as will the other parameters. The expected financial losses for individual h are:

$$p_1 \, (\alpha ME^h + \beta W^h).$$

Let us define quite arbitrarily CV_p^h by:

$$CV_p^h = CV^h - p_1 \left\{ \alpha ME^h + \beta W^h \right\}.$$

We thus have:

$$TSC = \sum_h \left\{ CV_p^h + p_1(\alpha ME^h + \beta W^h) \right.$$

$$+ p_1 \left[(1-\alpha)ME^h + (1-\beta)W^h \right]$$

$$\left. + p_2 p_1 \sum_{t=N^h}^{\overline{N}^h} \frac{NC^h}{(1+r)^{t-N^h}} \right\}$$

$$= \sum_h \left\{ CV_p^h + p_1 \left[ME^h + W^h \right] \right.$$

$$\left. + p_2 p_1 \sum_{t=N^h}^{\overline{N}^h} \frac{NC^h}{(1+r)^{t-N^h}} \right\}$$

The first term within the brackets can be interpreted as the psychic cost of the exposure to a greater risk; the second term the expected financial loss due to the disease and the third term the financial loss to the rest of the society as a result of a premature death. There are in general no particular problems involved in calculating the last two terms but there are grave estimation problems associated with the first term (see Chapters VI and VII), and reliable estimates may not be possible. If that is the case it is therefore not possible to estimate the total social cost of greater risks. In some cases when the probabliities p_1 and p_2 are small, the psychic cost may be small and can therefore be ignored.

Appendix C

POLLUTION AND LAND VALUES

This Appendix, through the use of a series of models discusses the effects of pollution or environmental quality on land values. The simplest model is that in which the difference in land values between two districts is determined solely by a difference in environmental quality. This model indicates the monetary value which a study of land values will yield for environmental quality if all individuals had identical preferences. The second model, in which there is an infinite number of districts with different environmental qualities, allows the assumption of identical preferences to be removed. The third model introduces transportation costs as an additional determinant of land values.

Two districts

Assume there are two districts, 1 and 2. The total supply of land in district i is S_i and the environmental quality is Y_i. We assume that there is a single homogeneous good produced in the two districts and that the production function can be written:

$$f(L, S, Y).$$

The production in district i is thus:

$$f(L_i, S_{if}, Y_i)$$

where L_i and S_{if} are labour and land allocated to production in district i. We assume that there is perfect competition on all markets and free entry, which means that profits are always zero and that the production function can be taken to be homogeneous of degree one in labour and land. We will neglect transportation costs for the moment so that the wage rate and the price on output will be the same in the two districts. Profit maximisation requires that:

$$pf_1^i = w$$

$$pf_2^i = r_i$$

where p is the price on output, w the wage rate and r_i the land rent in district i. f_j^i denotes the value in district i of the partial derivative of f with respect to argument number j. The marginal willingness-to-pay for an improvement of environmental quality in district i is given by:

(1) pf_3^1

i.e. the effect on profit from an improvement in environmental quality. Such a change will, however, lead to a re-adjustment of all prices. As the general equilibrium model determines only relative prices we can assume p = 1. The new equilibrium will be reached when all excess profits have disappeared, i.e. when (with π i denoting profit in district i):

$$\frac{d\pi_1}{dY_1} = f_3^1 + (f_1^1 - w)\frac{dL_1}{dY_1} + (f_2^1 - r_1)\frac{dS_{1f}}{dY_1} - L_1\frac{dw}{dY_1} - S_{1f}\frac{dr_1}{dY_1} = 0$$

$$\frac{d\pi_2}{dY_1} = (f_1^2 - w)\frac{dL_2}{dY_1} + (f_2^2 - r_2)\frac{dS_{2f}}{dY_1} - L_2\frac{dw}{dY_1} - S_{2f}\frac{dr_2}{dY_1} = 0$$

If we eliminate $\frac{dw}{dY_1}$ we get:

$$(2) \quad f_3^1 = S_{1f}\frac{dr_1}{dY_1} - \frac{L_1}{L_2}S_{2f}\frac{dr_2}{dY_1} = r_1 S_{1f}\left(\frac{1}{r_1}\frac{dr_1}{dy_1} - \frac{L_1}{S_{1f}}\frac{S_{2f}}{L_2}\frac{r_2}{r_1}\frac{1}{r_2}\frac{dr_2}{dY_1}\right)$$

If f is of Cobb-Douglas type, (2) can be simplified into

$$(3) \quad f_3^1 = r_1 S_{1f}\left(\frac{1}{r_1}\frac{dr_1}{dY_1} - \frac{1}{r_2}\frac{dr_2}{dY_1}\right).$$

The marginal damage of environmental deterioration on production can be calculated from information about the effects on land-values.

Let us now turn to the consumers. Firstly, we assume that the assumption of weak complementarity is valid, which seems reasonable. This means that a consumer living in one district does not care about environmental quality in the other district. Secondly, we assume that plots of land in the two districts are perfect substitutes and that the marginal rate of substitution is determined only by the environmental quality in the two districts. This means that the utility function of consumer h can be written:

$$(4) \quad U^h(C^h, L^h, B_1^h(Y_1)S_1^h + B_2^h(Y_2)S_2^h)$$

where C^h is the net demand of consumer h for the produced good, L^h his net supply of labour and S_1^h his net demand for land in district i.

Utility maximisation leads to the following conditions (subscripts denote partial derivatives and α^h is a Lagrangian multiplier):

$$(5)\quad \begin{cases} U_1^h = \alpha^h \\ U_2^h = \alpha^h w \\ U_3^h B_1^h \leqq \alpha^h r_1 \\ U_3^h B_2^h \leqq \alpha^h r_2 \end{cases}$$

Consider an individual h who is living in district 1. For him it is true that:

$$U_3^h B_1^h = \alpha^h r_1$$

$$U_3^h B_2^h = \alpha^h r_2$$

His marginal willingness-to-pay for a quality improvement in district 1 (or his demand price for Y_1) is given by(1):

$$(6)\quad \underline{\delta}^h = \frac{U_3^h}{U_1^h} B_1^{h'} \; S_1^h$$

where $B_1^{h'}$ is the derivative of B_1^h.

From (5) we see that:

$$\delta^h = \frac{U_3^h}{U_1^h} B_1^{h'} \; S_1^h = r_1 \; S_1^h \; \frac{B_1^{h'}}{B_1^h}.$$

Assume now that consumer h is a marginal individual, who is indifferent as to which district he inhabits. Accordingly, his marginal rate of substitution between land in the two districts must be equal to the relative land rent; that is,

$$\frac{B_1^h (Y_2)}{B_2^h (Y_2)} = \frac{r_1}{r_2}$$

Differentiating this with respect to Y_1, we have:

$$\frac{B_1^{h'}}{B_1^h} = \frac{1}{r_1} \frac{dr_1}{dy_1} - \frac{1}{r_2} \frac{dr_2}{dy_1}.$$

The demand price can thus be expressed as:

$$(7)\quad \delta^h = r_1 \; S_1^h \left\{ \frac{1}{r_1} \frac{dr_1}{dy_1} - \frac{1}{r_2} \frac{dr_2}{dy_1} \right\}$$

If all individuals are marginal, i.e. if they all have the same marginal rate of substitution between land in the two districts, the total marginal willingness-to-pay by consumers is given by:

$$\sum_h \delta \; h = \sum_h r_1 \; S_1^h \left\{ \frac{1}{r_1} \frac{dr_1}{dy_1} - \frac{1}{r_2} \frac{dr_2}{dy_1} \right\}$$

---

1) See Appendix A

where we sum over all individuals in district 1. If to this we add the marginal willingness-to-pay of the firm, we get the total marginal willingness-to-pay:

$$(8) \quad \zeta = \sum_h r_1 \, S_1^h \left\{ \frac{1}{r_1} \frac{dr_1}{dy_1} - \frac{1}{r_2} \frac{dr_2}{dy_1} \right\} + r_1 \, S_{1f} \left(\frac{1}{r_1} \frac{dr_1}{dy_1} - \frac{1}{r_2} \frac{dr_2}{dy_1} \right)$$

$$= \left(\frac{1}{r_1} \frac{dr_1}{dy_1} - \frac{1}{r_2} \frac{dr_2}{dy_1} \right) r_1 \left(\Sigma \, S_1^h + S_{1f} \right) =$$

$$r_1 \, S_1 \left(\frac{1}{r_1} \frac{dr_1}{dy_1} - \frac{1}{r_2} \frac{dr_2}{dy_1} \right).$$

The marginal willingness-to-pay for a quality improvement in district 1 can thus be calculated if one knows the effects on land values in the two districts; it is equal to the total value of land in district 1 multiplied by the difference between the relative change in land values in the two districts.

This discussion is based upon the following assumptions:

a) Perfect markets (which include the perfect perception of environmental changes) exist.

b) All individuals have the same preferences for environmental changes. If this assumption is not valid the relative land values will not represent the marginal rate of substitution between land in the two districts and it would be purely accidental if formula (8) gave the correct answer.

c) The production function is of the Cobb-Douglas type. If, however, production is not affected by changes in environmental quality this assumption is, of course, irrelevant.

d) There are no transportation costs.

Assumptions (a), (b) and (d) are probably the strongest ones and we will see to what extent it is possible to weaken them. It is obvious that we cannot abstain from assumption (a). In the next section we will look at assumption (b). Finally, in the last section transportation costs will be considered to see what changes occur in the results.

A Continuum of Districts

Let us assume that we can distinguish among different areas through use of an air quality variable Y, ranging in value between Y' and Y". Let t(Y) be the total area with air quality Y. The total area of the whole economy is then:

$$\int_{Y'}^{Y''} t(Y)dY.$$

Assume that there is a continuum of consumers represented by the interval [0,1], each indexed by h. Each consumer has a utility function:

$$U^h(C^h, L^h, S^h, Y)$$

where C^h is the consumption of consumer h, L^h his supply of labour, S^h his demand for land. The budget restriction is:

$$pC^h + r(Y)S^h = wL^h$$

where p is the price of consumption goods, w the wage rate and r(Y) the land rent as a function of air quality. We assume that p and w are independent of the air quality which means that transportation costs do not matter.

Utility maximisation yields:

$$(9) \begin{cases} U_1^h - \alpha p = 0 \\ U_2^h + \alpha w = 0 \\ U_3^h - \alpha r(Y) = 0 \\ U_4^h - \alpha r'(Y)S^h = 0. \end{cases}$$

This yields:

$$(10)\ r'(Y) = p\frac{U_4^h}{S^h U_1^h}$$

Let A(Y) be the set of consumers who demand land in areas where the air quality is Y. The total demand for land in such areas by consumers is then:

$$\int_{A(Y)} S^h dh$$

Assume now that the production activities in the economy can be described by the following production function:

$$x = f(L,S,Y)$$

where x is output, L the input of labour and S the input of land. For each Y, profit maximisation yields:

$$pf_1 - w = 0$$

$$pf_2 - r = 0.$$

The localisation of the plants is governed by:

$$pf_3 - r'(Y)S = 0.$$

The total land demanded for production and housing in areas with air quality Y is then:

$$S(Y) + \int_{A(Y)} S^h dh$$

Market clearance requires that this demand be equal to t(Y). Market equilibrium conditions for the other markets are simply:

$$\int_{Y'}^{Y''} x(Y)dY = \int_{0}^{1} C^h dh$$

$$\int_{Y'}^{Y''} L(Y)dY \quad \int_{0}^{1} L^h dh$$

The marginal willingness-to-pay for a quality improvement in areas with quality Y is the sum of the individual marginal willingness-to-pay, i.e.:

$$(11)\quad pf_3 + \int_{A(Y)} p\frac{U_4^h}{U_1^h}\, dh.$$

But according to our equilibrium conditions this can be written as:

$$S(Y)r'(Y) + \int_{A(Y)} r'(Y)D^h\, dh$$

$$(12)\quad = r'(Y)\left\{S(Y) + \int_{A(Y)} S^h dh\right\} = r'(Y)t(Y).$$

If the land value gradient r(Y) is known, it is possible to compute the total marginal willingness-to-pay for quality improvements by looking at the marginal increase in the total land value in a sector. Because we have a continuum of districts, a change in quality in one district will be so small that it will not change the land rents in other districts.

The term $\frac{1}{r_2}\frac{dr_2}{dy_1}$ in (8) is thus zero and (12) corresponds directly to (8). The assumption of a continuum of districts means that there are a sufficient number of districts to enable individuals within one district to constitute an approximately homogeneous group. The idea is simply that, if there are many districts with different environmental qualities, all individuals in one district can be regarded as marginal, because there exists another district to which the individuals would move if environmental quality were to change.

Transportation costs

Let us consider the situation where we have an urban centre in which all production activities take place. Environmental quality will in general vary within the area around this centre; for simplicity we assume that there is a function:

$$(13)\quad Y = \varphi(t)$$

relating the environmental quality of an area to the distance t from the centre.

A household living distance t from the centre has to pay transportation costs:

$$(14)\quad T = \varphi(t)$$

where φ is an increasing function.

The household must decide where to buy land and in so doing it must make a trade-off between transportation costs, land values and environmental quality. This decision problem can be formulated:

$$\max U(C, L, S, Y)$$

so that $wL = pC + r(t)S + \varphi(t)$

where $Y = \varphi(t)$

The necessary conditions for a maximum are (with α a Lagrangian multiplier):

$$(15)\quad \begin{cases} U_1 - \alpha p = 0 \\ U_2 + \alpha w = 0_1 \\ U_3 - rs = 0 \end{cases}$$

$$(16)\quad U_4\, \varphi' - \alpha(Sr' + \psi) = 0.$$

In equilibrium, the consumers must be indifferent to different localities and (16) must therefore be an identity in t. Eliminating α yields:

$$(17)\quad \delta = p\,\frac{U_h}{U_1} = \frac{1}{\varphi'}\left\{\varphi' + Sr'\right\}$$

The first term in the brackets on the right-hand side gives the increase in transportation costs if the household moves to units further from the centre. The second term gives the change in land rents resulting from such a move. The bracket thus gives the local change in cost ψ' is the corresponding change in environmental quality and the right-hand side is then an expression of the change in cost per unit change in environmental quality. The left-hand side is the marginal willingness-to-pay for an increase in environmental quality.

Let us assume now that we have estimated land rents as a function of distance from the centre and of environmental quality:

$$(18)\quad r = \tilde{r}(t, Y).$$

What kind of conclusions can be drawn from this estimate?

$\tilde{r}(t, Y)$ must satisfy:

$r(t) \equiv \tilde{r}[t_1\ \varphi(t)]$ which entails:

$$(19)\quad r' = \frac{\partial\tilde{r}}{\partial t} + \frac{\partial\tilde{r}}{\partial y}\,\varphi'.$$

Moreover, if Y is held constant, it is seen from (16) that:

$$(20)\quad S\frac{\partial\tilde{r}}{\partial t} = -\psi', \text{ and so:}$$

$$Sr' = \psi' + S\frac{\partial\tilde{r}}{\partial y}\,\varphi' \text{ and}$$

$$(21)\quad \delta = \frac{1}{\varphi'}(\psi' + Sr') = S\frac{\partial\tilde{r}}{\partial y}$$

With knowledge of the function $\tilde{\tilde{r}}(t,Y)$ it is thus possible to compute the marginal willingness-to-pay.

Appendix D

The derivation of formula (5-2) is as follows:

Let E be the total national expenditure on materials replacement, for example, and let the country be divided into two areas, a 'clean' area c, and a 'polluted' area p. Then:

(1) $E = n_c e_c + n_p e_p$

where e_i is the unit expenditure of replacement (per person, household) in area i and n_i is the number of units in area i. If the difference between e_p and e_c is due entirely to pollution then the total national cost of pollution is:

(2) $C = (n_c e_c + n_p e_p) - (n_c e_c + n_p e_c) = n_p(e_p - e_c)$

e_p and e_c are frequently unknown but the ratio F between them can often be easily estimated. Hence substituting Fe_c for e_p yields:

$$E = n_c e_c + n_p F e_c = (n_c + n_p F)e_c \text{ or:}$$

$$e_c = \frac{E}{(n_c + n_p F)}.$$

Using this result yields:

$$C = n_p(Fe_c - e_c)$$

$$= \frac{n_p E}{(n_c + n_p F)}(F - 1)$$

$$= \frac{E(F - 1)}{(F + \frac{n_c}{n_p})}$$

$$= \frac{E(F - 1)}{(F + R)}$$

where R is the ratio of the number of units in the 'clean' area to those in 'polluted' areas.

BIBLIOGRAPHY

The large number of articles which treat the many aspects of monetary environmental damage estimates prevents this bibliography from being an exhaustive one. The list presented below attempts to cover all of the important areas of interest, but it is not a complete or definitive list. Moreover, as the subject of titles chosen was influenced by their accessability to the authors, it is probable that many important titles are not among those presented below. The presence or absence of a title here implies no judgement on the value of its content.

In an effort to give the reader or future researcher some information about the contents of the works listed below, the following code was adopted to indicate their content.

A: Estimates of damage to materials.
B: Bibliographic resource.
D: Estimation of demand curves from individual behaviour.
E: Economic theory: environmental economics; welfare economics and cost-benefit analysis.
F: Estimates of financial losses.
G: General damage estimate studies.
H: Estimates of health damage, including annoyance.
L: Litigation as a methodology.
M: Methodology as related to economic theory.
P: Market reactions and property value methods.
Q: Questionnaire methods.
R: Estimation of recreation and natural resources damage.
U: Illustrations of damage function use.
V: Estimates of damage to vegetation and fauna.

R — Abel, Fred H. and Dennis P. Tihansky (1973) Methods and Problems of Estimating Water Quality Benefits and Costs, presented at the Society of Government Economists, Third Annual Conference, Washington D.C.

H, Q, P — Alexandre, A. and J. Ph. Barde (1973) Le Temps du Bruit, Flammarion, Paris.

P Anderson Jr., Robert J. and Thomas D. Crocker (1971) "Air Pollution and Residential Property Values", The Review of Economics and Statistics, pages 171-180.

G, R, H, V Appalachian Resources Project (1973) Statement on Benefit/ Cost Evaluation of Strip Mining in Appalachia, prepared for United States House of Representatives, Committee on Interior and Insular Affairs, The University of Tennessee, Knoxville, Tennessee.

F, G, H, U Argonne National Laboratory (1973) A Study of Social Costs for Alternate Means of Electrical Power Generation for 1980 and 1990, Argonne, Illinois.

E Arrow, K.J. and Fisher, A.C. (1974) Environmental Preservation, and Uncertainty, and Irreversibility, Quarterly Journal of Economics, May 1974, 88p, 312-19.

G, F, H, V, A, Q, P, B Barrett, Larry B. and Thomas E. Waddell (1973) Cost of Air Pollution Damage: A Status Report, United States Environmental Protection Agency, National Environmental Research Center, Research Triangle Park, North Carolina.

E Baumol, William J., (1969) "On the Discount Rate for Public Projects", The Analysis and Evaluation of Public Expenditures: The PPB System, Vol. 1., The Joint Economic Committee, Congress of the United States.

M, R Bender, M. et al (1974) Index of the Composite Environment (ICE): A Basis for Evaluating Environmental Effects of Electric Power Generating Plants in Response to NEPA, Oak Ridge National Laboratory, Oak Ridge, Tennessee.

V, B Benedict, H.M., C.J. Miller and R.E. Olson (1971) Economic Impact of Air Pollutants on Plants in the United States, Stanford Research Institute, Menlo Park, California.

E, M Bishop, John and Charles Cicchetti (1973) Some Institutional and Conceptual Thoughts on the Measurement of Indirect and Intangible Benefits and Costs, presented at an Environmental Protection Agency Symposium on Cost-Benefit Analysis and Pollution Control Policy, Annapolis, Maryland.

M Bohm, P. (1971) "An Approach to the Problem of Estimating Demand for Public Goods", Swedish Journal of Economics, p. 55.

M Bohm, Peter (1972) "An Approach to the Problem of Estimating Demand for Public Goods: An Experiment", European Economic Review, 1, p. 111.

R, G Bohm, Robert A., James H. Lord and David A. Patterson (1972) Market Imperfections, Social Costs of Strip Mining and Policy Alternatives, presented at the Southern Regional Science Association Meetings, 14th-15th April, 1972, Williamsburg, Virginia.

A, Q Booz, Allen and Hamilton, Inc. (1970) Study to Determine Residential Soiling Costs of Particulate Air Pollution, National Air Pollution Control Administration, Raleigh, North Carolina.

V, R Q, D Brown Jr., Gardner Mallard and Judd Hammack (1973) "Dynamic Economic Management of Migratory Waterfowl", The Review of Economics and Statistics, IV(1), pp. 73-82.

R, Q, D Brown, William G., Ajmer Singh and Emery N. Castle (1964) An Economic Evaluation of the Oregon Salmon and Steelhead Sport Fishery, Agricultural Experiment Station, Oregon State University, Corvallis, Oregon.

R, D Cesario, F.J., "A Method for Estimating Recreation Benefits", Appendix E in Resource Management in the Great Lakes Basin, Butrico et al editors, Heath Lexington Books, D.C. Heath and Company, Lexington, Massachusetts.

R, D Cicchetti, C.J. and A.C. Fisher, "Economic Values of Environmental Resources: Some Comments by Economists on the Supreme Court Decision in the Mineral King Case", unpublished manuscript.

R, D Q Cicchetti, Charles J., Joseph J. Seneca and Paul Davidson (1969) The Demand and Supply of Outdoor Recreation, Bureau of Economic Research, Rutgers - the State University, New Brunswick, New Jersey.

R, D Cicchetti, C.J., A.C. Fisher and V.K. Smith (1972) "An Econometric Model for the Valuation of Environmental Resources", presented at Econometric Society Meetings, Toronto, 27th-30th December, 1972.

R, D Clawson, M. and J.L. Knetsch (1971) Economics of Outdoor Recreation, Resources for the Future, Inc., the Johns Hopkins Press, Baltimore.

F, G, H, U Commission on Natural Resources, National Academy of Sciences, National Academy of Engineering, National Research Council (1975). Air Quality and Stationary Source Emission Control, prepared for the Committee on Public Works, United States Senate, Serial No. 94-4, United States Government Printing Office, Washington.

H, Q, P, U Commission on the Third London Airport (1971) Report, HMSO, London.

G, F, H, A Committee on Air Pollution (1954) Report, Her Majesty's Stationery Office, London.

R, D Common, M.S. (1973) "A Note on the Use of the Clawson Method for the Evaluation of Recreation Site Benefits", Regional Studies, 7, pp. 401-406.

M, R Coomber, Nicholas H. and Asit K. Biswas (1972) Evaluation of Environmental Intangibles - Review of Techniques, Ecological Systems Branch, Research Co-ordination Directorate, Policy, Planning and Research Service, Ottawa, Canada.

F, H, Q, P Copley International Corporation (1971) A Study of the Social and Economic Impacts of Odors, Phase II, Final Report Prepared for the Environmental Protection Agency, La Jolla, California.

M, P Crocker, Thomas (1971) "Urban Air Pollution Damage Functions: Theory and Measurements", University of California, Riverside.

M, P Crocker, Thomas D. (1972) "The Willingness-to-Pay for Air Pollution Control", Proceedings: AIME Council of Economics, pp. 45-54.

R, P Darling, Arthur H. (1973) "Measuring Benefits Generated by Urban Water Parks", Land Economics, XLIX(1), pp. 22-34.

E, H, Q, P Dasgupta, A.K. and D.W. Pearce (1972) Cost-Benefit Analysis: Theory and Practice, MacMillan, London.

R, D Davidson, P., F.G. Adams and J. Seneca (1966) "The Social Value of Water Recreational Facilities Resulting from an Improvement in Water Quality: The Delaware Estuary", in A. Kneese and S.C. Smith, editors, Water Research, Resources for the Future, Inc., Washington D.C., Johns Hopkins Press.

M Dehoz, Pierre (August 1974) "Analyse Théorique et Méthodes de Mesure des Avantages Collectifs de la Qualité de l'eau d'une Rivière". Document de Travail CB-V9, Center for Operations Research and Econometrics (C.O.R.E.), Louvain, Belgium.

A, F, G, H, U Denevers, Noel (1975) The Application of Cost-Benefit Analysis to Air Pollution Control Regulation. A General Discussion, with Application to the Example of Sulfur-Dioxide Control at the Huntington and Energy Power Plants submitted to the Utah Air Conservation Committee by Utah Power and Light Company, Salt Lake City.

R Deportes, J.P. (1972) Problemes économiques liés a l'assainissement du lac d'Annecy, Institut National de la Recherche Agronomique, Station d'hydrobiologie Lacustre, Thenon, France.

M, R Emmett (1974) Evaluation of Intangibles: State of the Art, Environmental Systems Branch, Planning and Finance Service, Department of the Environment, Ottawa, Canada.

F, R, V Enk, Gordon A., Project Director (1974) Assessing the Social Impacts of Oil Spills, Proceedings of an Invitational Symposium co-sponsored by the Institute for Man and Science and the United States Environmental Protection Agency, 25th-28th September, 1973, Rensselaerville, New York.

E Feldstein, M.S. (1969) "The Social Time Preference Discount Rate in Cost-Benefit Analysis", Economic Journal, 74, p. 360.

F, A, B Fink, F.W., F.H. Buttner and W.K. Boyd (1971) "Final Report on Technical-Economic Evaluation of Air Pollution Corrosion Costs on Metals in the United States" to Air Pollution Control Office, Environmental Protection Agency, Battelle Memorial Institute, Colombus, Ohio.

R, D Fisher, A.C. and J.V. Krutilla, The Economics of Environmental Resource Uses: Theory and Applications, to be published.

E Fisher, Anthony C. and John V. Krutilla (1973) Valuing Long Run Ecological Consequences and Irreversibilities presented at an Environmental Protection Agency Symposium on Cost-Benefit Analysis and Pollution Control Policy, Annapolis, Maryland.

R, D, U Fisher, Anthony C., John V. Krutilla and Charles J. Cicchetti (1972) "The Economics of Environmental Preservation: A Theoretical and Empirical Analysis", American Economic Review, LXII(4), pp. 605-619.

H, P Flowerdew, A.D.J., "The Cost of Airport Noise", The Statistician, 21(1), pp. 31-46.

M, P Freeman III, A. Myrick (1974) "On Estimating Air Pollution Control Benefits from Land Value Studies", Journal of Environmental Economics and Management, 1, pp. 74-83.

M, F, D, Q, P Freeman III, A. Myrick (1973) A Survey of the Techniques for Measuring the Benefits of Water Quality Improvement presented at an Environmental Protection Agency Symposium on Cost-Benefit Analysis and Pollution Control Policy, Annapolis, Maryland.

E Freeman III, A. Myrick, Robert H. Haveman and Allen V. Kneese (1973) The Economics of Environmental Policy, John Wiley and Sons, Inc., New York.

V, R Gaudet, Jean-Louis (1973) Economic Issues and Opportunities Facing Europe in the Field of Sport Fisheries, European Inland Fisheries Advisory Commission, Food and Agriculture Organisation of the United Nations, Rome.

E Graaff, J. de V. (1957) Theoretical Welfare Economics, Cambridge University Press.

E, R Hammond, R.J. (1959) Benefit-Cost Analysis and Water Pollution Control, Food Research Institute, Stanford University, Palo Alto, California.

E, G, F, H Haveman, Robert H. (1975) "On Estimating Environmental Damage: A Survey of Recent Research in the United States" in Environmental Damage Costs, OECD, Paris.

E Haveman, Robert H. and Burton A. Weisbrod (1973) The Concept of Benefits in Benefit-Cost Analysis with Emphasis on Water Pollution Control Activities, presented at an Environmental Protection Agency Symposium on Cost-Benefit Analysis and Water Pollution Policy, Annapolis, Maryland.

F, H, V Headly, J.C. and J.N. Lewis (1967) The Pesticide Problem: An Economic Approach to Public Policy, Resources for the Future, The Johns Hopkins Press, Baltimore.

V, R, D Helliwell, D.R. (1969) "Valuation of Wildlife Resources", Regional Studies, 3, pp. 41-47.

E Henry, C. (1974) Investment Decisions Under Uncertainty: The "Irreversibility Effect" The American Economic Review, December 1974, pp. 1006-1012.

H, L Hereford, Peggy G. (1973) Recap of Lawsuits, Court Decisions and California State Legislation - Impact upon Department of Airports and Its Rôle as Set Forth by City Charter to Accommodate Air Commerce and Navigation, Los Angeles Department of Airports, Los Angeles.

E Hicks, John R., "The Generalised Theory of Consumers' Surplus", Review of Economic Studies, XIII, 34, p. 68.

E Hicks, John R., "The Four Consumers' Surpluses", Review of Economic Studies, XI(1), p. 31.

E Hicks, John R., "The Rehabilitation of Consumers' Surplus", Review of Economic Studies, VIII(2), p. 108.

H, Q Hoinville, G. and P. Prescott-Clarke (1972) Traffic Disturbance and Amenity Values, Social and Community Planning Research (SCPR), London.

G, F, H, V, A, Q, P Jansen, H.M.A. G.J. van der Meer, J.B. Opschoor and J.H.A. Stapel (1974) An Estimate of Damage Caused by Air Pollution in the Netherlands in 1970, English Summary, Institute for Environmental Problems, Free University of Amsterdam.

G, F, H, V, A, Q, P Jansen, H.M.A., G.J. van der Meer, J.B. Opschoor and J.H.A. Stapel (1974) Een Raming van de Schade Door Luchtverontreiniging in Nederland in 1970, Institut voor Milieu-vraagstukken, Vrije Universiteit, Amsterdam.

V, B Jonsson, Bengt and Rolf Sundberg (1972) Has the Acidification by Atmospheric Pollution Caused a Growth Reduction in Swedish Forests? Institutionen för Skogsproduktion, Skogsnögokolan, Stockholm.

E, R Jordening, David L. and James K. Allwood (1973) Research Needs and Priorities: Pollution Control Benefits and Costs with Emphasis on Water, A Report to the Office of Research and Monitoring, United States Environmental Protection Agency, Development Planning and Research Associates, Manhattan, Kansas.

E, U Jürgensen, Harald (1973) Study to Determine the "Social Costs" of Environmental Pollution, communicated to the European Communities, Hamburg.

R, D Kavanagh, N.J. and J.G. Gibson (1971) "Measurement of Fishing Benefits on the River Trent", from a Symposium on the Trent Research Programme, The University of Nottingham.

R Keilani, W.M., R.H. Peters and P.J. Reynolds (1974) A Water Quality Economic Index, presented at the Ninth Canadian Symposium, on Water Pollution Research, the University of Western Ontario, London, Ontario.

H Kidner R. and K. Richards (1974) "Compensation to Dependants of Accident Victims", The Economic Journal, pp. 130-142.

E Kneese, Allen V., Robert V. Ayres and Ralph C. d'Arge, (1972) Economics and the Environment: A Materials Balance Approach, Resources for the Future, Inc., Washington D.C.

E Kneese, Allen V. and Blair T. Bower, eds. (1972) Environmental Quality Analysis: Theory and Method in the Social Sciences, Johns Hopkins University Press, Baltimore.

E Kneese, Allen V. and Blair T. Bower (1968) Managing Water Quality: Economics, Technology and Institutions, Johns Hopkins University Press.

E Kolm, Serge-Christophe (1975) "Connaissance des Coûts et Valeurs d'Environnement" in Coûts des Domages Causés à l'Environnement, OECD, Paris.

R, D, U Krutilla, J. and C. Cicchetti (1972) "Evaluating Benefits of Environmental Resources with Special Application to the Hells Canyon", National Resources Journal 12(1), pp. 1-29

M Kurz, Modercai (October 1973) "Experimental Approach to the Determination of the Demand for Public Goods", NSP Grant Series, Working Paper III, Institute for Mathematical Studies in the Social Sciences, Stanford University, Palo Alto, California.

V, Q, Landau, Emanual (1971) Statistical Aspects of Air Pollution as it Applies to Agriculture, submitted at the 1971 Joint Meeting of the Statistical Societies, Fort Collins, Colorado.

H, Q, D, P Lassière, A. and P. Bowers (1972) Studies on the Social Costs of Urban Road Transport (Noise and Pollution), Report of the Eighteenth Round Table on Transport Economics, European Conference of Ministers of Transport, Paris.

H Lave, L.B. and E.P. Seskin (1970) "Air Pollution and Human Health", Science, 169, pp. 723-733.

M Lind, R.C. (1966) The Nature of Flood Control Benefits and the Economics of Flood Protection, Stanford University, Palo Alto, California.

M Lind, Robert C. (1973) "Spatial Equilibrium, The Theory of Rents and the Measurements of Benefits from Public Programs", Quarterly Journal of Economics, 87 (2), p. 188.

V Linzon, Samuel N. (1971) "Economic Effects on Sulphur Dioxide on Forest Growth", Journal of the Air Pollution Control Association, 21 (2), pp. 81-86.

E, M, Q Mäler, Karl-Göran (1975) "Damage Functions and Their Estimation: A Theoretical Survey", in Environmental Damage Costs, OECD, Paris.

E, M Mäler, Karl-Göran (1974) Environmental Economics: A Theoretical Inquiry, Johns Hopkins Press, Baltimore.

M Mäler, Karl-Göran (1971) "A Method of Estimating Social Benefits from Pollution Control", Swedish Journal of Economics, 73(1).

E Margolis, Julius (1969) "Shadow Prices for Incorrect or Non-existant Market Values", The Analysis and Evaluation of Public Expenditures: The PPB System, The Joint Economic Committee, Congress of the United States.

V, Q, F Millecan, Arthur A. (1971) A Survey and Assessment of Air Pollution Damage to California Vegetation in 1970, California Department of Agriculture, Sacramento, California.

E Misban, Ezra J. (1971) "Evaluation of Life and Limb: A Theoretical Approach", The Journal of Political Economy, 79(4), pp. 6-7.

E Misban, Ezra J. (1966) "A Survey of Welfare Economics, 1939-1959" in Surveys of Economic Theory Vol. I., MacMillan, London.

G, F, R, H, B Morgan, M. Granger, Barbara R. Barkovich and Alan K. Meier (1973) "The Social Costs of Producing Electric Power from Coal: A First Order Calculation", *Proceedings of the IEEE*.

L, U Moroshima, Akio (1975) "Damage Functions from the Legal Perspective: Japanese Legal Experiences in Pollution Damage", in *Environmental Damage Costs*, OECD, Paris.

E, G Muraro, Gilberto (1975) "Estimate of the Economic Damage Caused by Pollution: The Italian Experience", in *Environmental Damage Costs*, OECD, Paris.

R, D, P The National Board of Water, Finland. *The Benefits and Costs of Water Pollution Control in Lohja District*.

G, F, A O'Connor, J.J. (1913) *The Economic Costs of the Smoke Nuisance to Pittsburgh*, Mellon Institute, Pittsburgh, Pa.

G, F, H, P Opschoor, Hans B. (1975) "Damage Functions, Some Theoretical and Practical Problems", in *Environmental Damage Costs*, OECD, Paris.

H, D, P Paul, M.E. (1971) "Can Aircraft Noise Nuisance be Measured in Money?", *Oxford Economic Papers*, pp. 297-322.

M, D, R Pearse, Peter H. (1968) "A New Approach to the Evaluation of Non-Priced Recreational Resources" *Land Economics, 64*, pp. 87-99.

H, Q Pendse, Dilip and J.B. Wyckoff (1974) "Scope for Valuation of Environmental Goods", *Land Economics L (1)*, p. 89

H, Q, P Plowden, S. (1970) *The Cost of Noise*, Metra Consulting Group Ltd, London.

E Prest, A.R. and Ralph Turvey (1966) "Cost-Benefit Analysis: A Survey", in *Surveys of Economic Theory, Vol. III*, MacMillan, London.

G, F, H, V, A Programmes Analysis Unit (1971) *An Economic and Technical Appraisal of Air Pollution in the United Kingdom*, Chilton, Didcot, Berkshire.

F, R Programmes Analysis Unit (1972) *The Environmental and Financial Consequences of Oil Pollution from Ships*, Main Report, Appendices 1, 2, 3 and 4, Report of Study No. VI submitted by the United Kingdom, Chilton, Didcot, Berkshire.

G, U RECAT Committee (1972) *Cumulative Regulatory Effects on the Cost of Automotive Transportation*, Final Report of the *ad hoc* Committee, prepared for the Office of Science and Technology, Washington D.C.

M, D, Reiling, S.D., K.C. Gibbs and H.H. Stoevener (1973) Economic
R Benefits from an Improvement in Water Quality, prepared for Office of Research and Monitoring, United States Environmental Protection Agency, Washington D.C.

H, D, Richards, E.J. (1971) "Noise and Society", Journal of the
P Royal Society of Arts, pp. 678-698.

G, F, Ridker, R.G. (1967) Economic Costs of Air Pollution, Praeger,
H, A, New York.
Q, P

G, V, Royal Ministry for Foreign Affairs and Royal Ministry of
A Agriculture (1971) Air Pollution Across National Boundaries: The Impact on the Environment of Sulphur in Air and Precipitation, Sweden's Case Study for the United Nations Conference on the Human Environment.

F, H Sagan, L.A. (1972) "Human Costs of Nuclear Power", Science, 177, pp. 487-493.

F, A, Salmon, R.L. (1970) Systems Analysis of the Effects of Air
B Pollution in Materials, Final Report for Economic Effects, Research Division, National Air Pollution Control Administration, Raleigh, North Carolina, published by Midwest Research Institute, Kansas City, Missouri.

E Samuelson, Paul A. (1954) "The Pure Theory of Public Expenditures", Review of Economics and Statistics, 36, pp. 387-389.

M Sandmo, Agnar, "Public Goods and the Technology of Consumption", Review of Economic Studies.

E Sen, A.K. (1967) "Isolation, Assurance and the Social Rate of Discount", Quarterly Journal of Economics, 81, p. 112.

R, D Seneca, Joseph J. (1969) "Water Recreation, Demand and Supply", Water Resources Research, 5(6), pp. 1170-1185.

F, A Spence, J.W. and F.H. Haynie (1972) Paint Technology and Air Pollution: A Survey and Economic Assessment, United States Environmental Protection Agency, Research Triangle Park, North Carolina.

M, P Spore, Robert Leo (1972) Property Value Differentials as a Measure of the Economic Costs of Air Pollution, Center for Air Environment Studies, the Pennsylvania State University, University Park, Pennsylvania.

H, Q, Starkie, D.N.M. (1974) Valuation of Disamenity in Transport
D Planning, A Report to the Secretary of State for the Environment, Department of Geography, University of Reading, United Kingdom.

H, Q, D, P Starkie, D.N.M. and D.J. Johnson (1973) "Losses of Residential Amenity: An Extended Cost Model", *Journal of Regional Studies, 7(2)*.

R, D Stevens, J. (1966) "Recreational Benefits from Water Pollution Control", *Water Resources Research, 2*.

F, A Stickney, P.B., W.J. Mueller and J.W. Spence (1971) "Pollution vs. Rubber", *Rubber Age*, pp. 45-51.

M Strotz, Robert N. (1966) *Economics of Urban Air Pollution*, Resources for the Future, Inc. Washington D.C.

M, P Strotz, R.H. (1966) *The Use of Land Rent Charges to Measure the Welfare Benefits of Land Improvement*, Resources for the Future, Inc.

G Systems Research Group, Damage Functions, EQM Working Paper No. 15, Toronto, Ontario.

D, E, H Thaler, Richard and Sherwin Rosen (1973) *The Value of Saving a Life: Evidence from the Labor Market*, Graduate School of Management, University of Rochester, Rochester, New York.

G, F, V, R Tihansky, Dennis P. (1973) *Estimating Economic Damages from Marine Pollution Along United States Shores*, United States Environmental Protection Agency, Washington D.C.

F, R, B Tihansky, Dennis P. (1973) *A Survey of Empirical Benefit Studies*, presented at an Environmental Protection Agency Symposium on Cost-Benefit Analysis and Pollution Control Policy, Annapolis, Maryland.

R, D Ullman, Edward L. and Donald J. Volk (1962) "An Operational Model for Predicting Reservoir Attendance and Benefits: Implications of a Location Approach to Water Recreation", *Papers of the Michigan Academy of Science, Arts and Letters*, *XLVII*, pp. 473-484.

E, M, F, R, B Unger, Samuel G., M. Jarvin Emerson and David L. Jordening (1973) State-of-Art Review: *Pollution Control Benefits and Costs with Emphasis on Water*, A Report to the Office of Research and Monitoring, United States Environmental Protection Agency, Development Planning and Research Associates, Manhattan, Kansas.

F, G, H, U United States Atomic Energy Commission (1974) *Comparative Risk-Cost-Benefit Study of Alternative Sources of Electrical Energy* WASH-1224, United States Government Printing Office, Washington, D.C.

H, L United States Department of Transportation (1972) Societal Costs of Motor Vehicle Accidents, Preliminary Report, National Highway Traffic Safety Administration, Washington D.C.

G, R, Q Vers Jr., Charles R. and Gary W. Sorenson (1972) Study of the Economic Effects of Changes in Air Quality, Oregon State University, Corvallis, Oregon.

R, D Vickerman, Roger (1972) "The Demand for Non-Work Travel", Journal of Transport Economics and Policy, VI(2), pp. 1-35.

R, D Vickerman, R.W. (1974) "A Demand Model for Leisure Travel", Environment and Planning A, 6, pp. 65-77.

R, D Vickerman, R.W. "The Evaluation of Benefits from Recreational Projects", submitted to Urban Studies.

G, F, H, V, A, Q, P, L Waddell, Thomas E. (1974) The Economic Damages of Air Pollution, Washington Environmental Research Center, Office of Research and Development, United States Environmental Protection Agency, Washington D.C.

E Weisbrod, B.A. (1968) "Income Redistribution Effects and Benefit-Cost Analysis", in Chase B.S. (ed.) Problems in Public Expenditure Analysis, p. 177, The Brookings Institute, Washington D.C.

R Whitman, Ira L. et al (1971) Design of an Environmental Evaluation System, Battelle Memorial Institute, Columbus, Ohio.

G, F, H, V, R, A, P, B Wyzga, R.E. (197) "A Survey of Environmental Damage Functions", in Environmental Damage Costs, OECD, Paris.

OECD SALES AGENTS
DEPOSITAIRES DES PUBLICATIONS DE L'OCDE

ARGENTINA – ARGENTINE
Carlos Hirsch S.R.L.,
Florida 165, BUENOS-AIRES.
☎ 33-1787-2391 Y 30-7122

AUSTRALIA – AUSTRALIE
International B.C.N. Library Suppliers Pty Ltd.,
161 Sturt St., South MELBOURNE, Vic. 3205.
☎ 69.7601
658 Pittwater Road, BROOKVALE NSW 2100.
☎ 938 2267

AUSTRIA – AUTRICHE
Gerold and Co., Graben 31, WIEN 1.
☎ 52.22.35

BELGIUM – BELGIQUE
Librairie des Sciences
Coudenberg 76-78, B 1000 BRUXELLES 1.
☎ 512-05-60

BRAZIL – BRESIL
Mestre Jou S.A., Rua Guaipá 518,
Caixa Postal 24090, 05089 SAO PAULO 10.
☎ 256-2746/262-1609
Rua Senador Dantas 19 s/205-6, RIO DE JANEIRO GB. ☎ 232-07. 32

CANADA
Information Canada
171 Slater, OTTAWA. KIA 0S9.
☎ (613) 992-9738

DENMARK – DANEMARK
Munksgaards Boghandel
Nørregade 6, 1165 KØBENHAVN K.
☎ (01) 12 69 70

FINLAND – FINLANDE
Akateeminen Kirjakauppa
Keskuskatu 1, 00100 HELSINKI 10. ☎ 625.901

FRANCE
Bureau des Publications de l'OCDE
2 rue André-Pascal, 75775 PARIS CEDEX 16.
☎ 524.81.67
Principaux correspondants :
13602 AIX-EN-PROVENCE : Librairie de l'Université. ☎ 26.18.08
38000 GRENOBLE : B. Arthaud. ☎ 87.25.11
31000 TOULOUSE : Privat. ☎ 21.09.26

GERMANY – ALLEMAGNE
Verlag Weltarchiv G.m.b.H.
D 2000 HAMBURG 36, Neuer Jungfernstieg 21
☎ 040-35-[illegible]2-500

GREECE – GRECE
Librairie Kauffmann, 28 rue du Stade,
ATHENES 132. ☎ 322.21.60

HONG-KONG
Government Information Services,
Sales of Publications Office,
1A Garden Road,
☎ H-252281-4

ICELAND – ISLANDE
Snaebjörn Jónsson and Co., h.f.,
Hafnarstræti 4 and 9, P.O.B. 1131,
REYKJAVIK. ☎ 13133/14281/11936

INDIA – INDE
Oxford Book and Stationery Co. :
NEW DELHI, Scindia House. ☎ 47388
CALCUTTA, 17 Park Street. ☎ 24083

IRELAND – IRLANDE
Eason and Son, 40 Lower O'Connell Street,
P.O.B. 42, DUBLIN 1. ☎ 74 39 35

ISRAEL
Emanuel Brown :
35 Allenby Road, TEL AVIV. ☎ 51049/54082
also at :
9, Shlomzion Hamalka Street, JERUSALEM.
☎ 234807
48 Nahlath Benjamin Street, TEL AVIV.
☎ 53276

ITALY – ITALIE
Libreria Commissionaria Sansoni :
Via Lamarmora 45, 50121 FIRENZE. ☎ 579751
Via Bartolini 29, 20155 MILANO. ☎ 365083
Sous-dépositaires :
Editrice e Libreria Herder,
Piazza Montecitorio 120, 00186 ROMA.
☎ 674628
Libreria Hoepli, Via Hoepli 5, 20121 MILANO.
☎ 865446
Libreria Lattes, Via Garibaldi 3, 10122 TORINO.
☎ 519274
La diffusione delle edizioni OCDE è inoltre assicurata dalle migliori librerie nelle città più importanti.

JAPAN – JAPON
OECD Publications Centre,
Akasaka Park Building,
2-3-4 Akasaka,
Minato-ku
TOKYO 107. ☎ 586-2016
Maruzen Company Ltd.,
6 Tori-Nichome Nihonbashi, TOKYO 103,
P.O.B. 5050, Tokyo International 100-31.
☎ 272-7211

LEBANON – LIBAN
Documenta Scientifica/Redico
Edison Building, Bliss Street,
P.O.Box 5641, BEIRUT. ☎ 354429 – 344425

THE NETHERLANDS – PAYS-BAS
W.P. Van Stockum
Buitenhof 36, DEN HAAG. ☎ 070-65.68.08

NEW ZEALAND – NOUVELLE-ZELANDE
The Publications Manager
Government Printing Office
Mulgrave Street (Private Bag)
WELLINGTON. ☎ 737-320
and Government Bookshops at
AUCKLAND (P.O.B. 5344). ☎ 32.919
CHRISTCHURCH (P.O.B. 1721). ☎ 50.331
HAMILTON (P.O.B. 857). ☎ 80.103
DUNEDIN (P.O.B. 1104). ☎ 78.294

NORWAY – NORVEGE
Johan Grundt Tanums Bokhandel,
Karl Johansgate 41/43, OSLO 1. ☎ 02-332980

PAKISTAN
Mirza Book Agency, 65 Shahrah Quaid-E-Azam,
LAHORE 3. ☎ 66839

PHILIPPINES
R.M. Garcia Publishing House,
903 Quezon Blvd. Ext., QUEZON CITY,
P.O. Box 1860 – MANILA. ☎ 99.98.47

PORTUGAL
Livraria Portugal,
Rua do Carmo 70-74. LISBOA 2. ☎ 360582/3

SPAIN – ESPAGNE
Libreria Mundi Prensa
Castelló 37, MADRID-1. ☎ 275.46.55
Libreria Bastinos
Pelayo, 52, BARCELONA 1. ☎ 222.06.00

SWEDEN – SUEDE
Fritzes Kungl. Hovbokhandel,
Fredsgatan 2, 11152 STOCKHOLM 16.
☎ 08/23 89 00

SWITZERLAND – SUISSE
Librairie Payot, 6 rue Grenus, 1211 GENEVE 11.
☎ 022-31.89.50

TAIWAN
Books and Scientific Supplies Services, Ltd.
P.O.B. 83, TAIPEI.

TURKEY – TURQUIE
Librairie Hachette,
469 Istiklal Caddesi,
Beyoglu, ISTANBUL. ☎ 44.94.70
et 14 E Ziya Gökalp Caddesi
ANKARA. ☎ 12.10.80

UNITED KINGDOM – ROYAUME-UNI
H.M. Stationery Office, P.O.B. 569, LONDON
SE1 9 NH, ☎ 01-928-6977, Ext. 410
or
49 High Holborn
LONDON WC1V 6HB (personal callers)
Branches at: EDINBURGH, BIRMINGHAM, BRISTOL, MANCHESTER, CARDIFF, BELFAST.

UNITED STATES OF AMERICA
OECD Publications Center, Suite 1207,
1750 Pennsylvania Ave, N.W.
WASHINGTON, D.C. 20006. ☎ (202)298-8755

VENEZUELA
Libreria del Este, Avda. F. Miranda 52,
Edificio Galipán, Aptdo. 60 337, CARACAS 106.
☎ 32 23 01/33 26 04/33 24 73

YUGOSLAVIA – YOUGOSLAVIE
Jugoslovenska Knjiga, Terazije 27, P.O.B. 36,
BEOGRAD. ☎ 621-992

Les commandes provenant de pays où l'OCDE n'a pas encore désigné de dépositaire peuvent être adressées à :
OCDE, Bureau des Publications, 2 rue André-Pascal, 75775 Paris CEDEX 16
Orders and inquiries from countries where sales agents have not yet been appointed may be sent to
OECD, Publications Office, 2 rue André-Pascal, 75775 Paris CEDEX 16

OECD PUBLICATIONS, 2, rue André-Pascal, 75775 Paris Cedex 16 - No. 35.267 1976

PRINTED IN FRANCE